SAXON MATH™
Intermediate 4

Reteaching Masters

Stephen Hake

A Harcourt Achieve Imprint

www.SaxonPublishers.com
1-800-284-7019

Introduction

Monitor student progress during daily instruction and evaluate after each Cumulative Test. Assign reteaching masters as necessary to reinforce mathematical concepts and skills.

Use Individual Test Analysis Form B from the *Assessment Guide* to identify those concepts that may be causing students difficulty. Complete a separate form after every Cumulative Test for each student who scored below 80%. Not all students will have mastered a new concept the first time it is assessed. However, if a student has not mastered a new concept after repeated practice and assessment, then reteaching is indicated.

Each reteaching master contains a concise summary of the new concept introduced in a lesson or investigation, including annotated examples. The lesson summary is followed by a set of practice items to support reteaching and remediation of the new concept.

ISBN-13: 978-1-6003-2524-3

ISBN-10:   1-6003-2524-6

© 2008 Harcourt Achieve Inc. and Stephen Hake.

All rights reserved. This book is intended for classroom use and is not for resale or distribution. Each blackline master in this book may be reproduced, with the copyright notice, without permission from the Publisher. Reproduction for an entire school or district is prohibited. No other part of this publication may be reproduced or transmitted in any form or by any means, electronic or mechanical, including photocopying, recording, taping, or any information storage and retrieval system, without permission in writing from the Publisher. Requests for permission should be mailed to: Paralegal Department, 6277 Sea Harbor Drive, Orlando, FL 32887.

Saxon is a trademark of Harcourt Achieve Inc.

Printed in the United States of America

3 4 5 6 7 8   862   15 14 13 12 11 10 09

Table of Contents

Reteaching

Reteaching 1 1
Reteaching 2 2
Reteaching 3 3
Reteaching 4 4
Reteaching 5 5
Reteaching 6 6
Reteaching 7 7
Reteaching 8 8
Reteaching 9 9
Reteaching 10 10
Reteaching Inv. 1 11
Reteaching 11 12
Reteaching 12 13
Reteaching 13 14
Reteaching 14 15
Reteaching 15 16
Reteaching 16 17
Reteaching 17 18
Reteaching 18 19
Reteaching 19 20
Reteaching 20 21
Reteaching Inv. 2 22
Reteaching 21 23
Reteaching 22 24
Reteaching 23 25
Reteaching 24 26
Reteaching 25 27
Reteaching 26 28
Reteaching 27 29
Reteaching 28 30
Reteaching 29 31

Table of Contents

Reteaching 30 32
Reteaching Inv. 3 33
Reteaching 31 34
Reteaching 32 35
Reteaching 33 36
Reteaching 34 37
Reteaching 35 38
Reteaching 36 39
Reteaching 37 40
Reteaching 38 41
Reteaching 39 42
Reteaching 40 43
Reteaching Inv. 4 44
Reteaching 41 45
Reteaching 42 46
Reteaching 43 47
Reteaching 44 48
Reteaching 45 49
Reteaching 46 50
Reteaching 47 51
Reteaching 48 52
Reteaching 49 53
Reteaching 50 54
Reteaching Inv. 5 55
Reteaching 51 56
Reteaching 52 57
Reteaching 53 58
Reteaching 54 59
Reteaching 55 60
Reteaching 56 61
Reteaching 57 62
Reteaching 58 63
Reteaching 59 64
Reteaching 60 65

Table of Contents

Reteaching Inv. 6 66
Reteaching 61 67
Reteaching 62 68
Reteaching 63 69
Reteaching 64 70
Reteaching 65 71
Reteaching 66 72
Reteaching 67 73
Reteaching 68 74
Reteaching 69 75
Reteaching 70 76
Reteaching Inv. 7 77
Reteaching 71 78
Reteaching 72 79
Reteaching 73 80
Reteaching 74 81
Reteaching 75 82
Reteaching 76 83
Reteaching 77 84
Reteaching 78 85
Reteaching 79 86
Reteaching 80 87
Reteaching Inv. 8 88
Reteaching 81 89
Reteaching 82 90
Reteaching 83 91
Reteaching 84 92
Reteaching 85 93
Reteaching 86 94
Reteaching 87 95
Reteaching 88 96
Reteaching 89 97
Reteaching 90 98
Reteaching Inv. 9 99

Reteaching 91	100
Reteaching 92	101
Reteaching 93	102
Reteaching 94	103
Reteaching 95	104
Reteaching 96	105
Reteaching 97	106
Reteaching 98	107
Reteaching 99	108
Reteaching 100	109
Reteaching Inv. 10	110
Reteaching 101	111
Reteaching 102	112
Reteaching 103	113
Reteaching 104	114
Reteaching 105	115
Reteaching 106	116
Reteaching 107	117
Reteaching 108	118
Reteaching 109	119
Reteaching 110	120
Reteaching Inv. 11	121
Reteaching 111	122
Reteaching 112	123
Reteaching 113	124
Reteaching 114	125
Reteaching 115	126
Reteaching 116	127
Reteaching 117	128
Reteaching 118	129
Reteaching 119	130
Reteaching 120	131
Reteaching Inv. 12	132
Answer Key	133

Name _____

Reteaching 1
Lesson 1

- **Review of Addition**

  - Added numbers are called **addends** and the answer is the **sum**.

    addend + addend = sum

    Example:

    ```
      5        addend
    + 2      + addend
      7        sum
    ```

  - **The Commutative Property of Addition** tells us that changing the order of the addends does not change the sum.

    $6 + 3 = 9$ $3 + 6 = 9$

  - **The Identity Property of Addition** tells us that when we add zero to a number, that number does not change.

    $7 + 0 = 7$ $0 + 2 = 2$

  - The expression $2 + 6 = 8$ is a **number sentence**.

  - "Some and some more" problems have an addition formula.

Formula	Problem
Some	6 volleyballs
+ Some more	+ 7 volleyballs
Total	13 volleyballs

  - To find a missing addend, we subtract the known addend from the sum.

    ```
      5          8          n          6
    + n        − 5        + 4        − 4
      8        n = 3        6        n = 2
    ```

Practice:

1. $5 + 4 = $ _____

2. $3 + 0 = $ _____

3. $1 + 3 + 7 = $ _____

4. Write two number sentences to show the commutative property of 3 and 8:

   _____ + _____ = _____

   _____ + _____ = _____

Find the missing addend.

5. $7 + n = 12$

   $n = $ _____

6. $n + 5 = 13$

   $n = $ _____

Saxon Math Intermediate 4

Name _____

Reteaching 2
Lesson 2

- # Missing Addends

  - To find a missing **addend**, we subtract the sum of the given addends from the given total.

    Example:
    ```
         7        7 + 6 = 13      sum of given addends
         n        18 − 13 = 5     subtract from total
        + 6       n = 5           missing addend
        ----
        18
    ```

  - Look for pairs of addends that can be added together to equal 10. These are "sets of 10."

    Sets of 10
9 + 1 = 10
8 + 2 = 10
7 + 3 = 10
6 + 4 = 10
5 + 5 = 10

Practice:

Find each missing addend.

1. $9 + 3 + n = 16$

   $9 + 3 = 12$

   $16 - 12 =$ _____

   $n =$ _____

2. $x + 5 + 4 = 16$

   $5 + 4 = 9$

   $16 - 9 =$ _____

   $n =$ _____

3. $7 + y + 4 + 8 = 25$

   $7 + 4 + 8 = 19$

   $25 - 19 =$ _____

   $y =$ _____

4. $6 + 5 + n + 9 + 2 + 7 = 34$

   $6 + 5 + 9 + 2 + 7 =$ _____

   $34 - 29 =$ _____

   $n =$ _____

Find sets of 10. Add.

5. $9 + 2 + 6 + 4 + 5 + 1 + 8 =$ _____

6. $5 + 4 + 7 + 3 + 9 + 2 + 1 + 1 =$ _____

7. $8 + 4 + 2 + 6 + 3 + 1 + 7 + 9 + 5 =$ _____

Name _____

Reteaching 3
Lesson 3

- **Sequences**
- **Digits**

Sequences
- **Counting numbers** have no end.
    1, 2, 3, 4, 5, ...
- A **sequence** is a counting pattern. It can go "up" or "down."
    5, 10, 15, 20, 25, ...
    20, 15, 10, 5, ...
- Subtract to find the **rule**.

    Example:
    9, 13, 17, ____, ____, ____, ...
    We can also look at the 4s row in the times
    table to find other numbers in the sequence.
    The rule for this sequence is count up by fours.

    13 17
    -9 -13
    4 4

    +4 +4 +4 +4 +4
    9, 13, 17, 21, 25, 29

Digits
- Digits are the numerals 0, 1, 2, 3, 4, 5, 6, 7, 8, and 9.
    471 has three digits.
    The last digit is 1.

Practice:

Write the rule and the next two numbers of each counting sequence.

1. 9, 8, 7, _____, _____, ...

    Rule: Count down by _____.

2. 2, 5, 8, 11, _____, _____, ...

    Rule: Count up by _____.

Find the missing number in each counting sequence.

3. 35, 30, 25, _____, 15 ...

4. 2, _____, 12, 17, 22, 27 ...

How many digits are in each number?

5. 108 _____

6. 5372 _____

What is the last digit of each number?

7. 214 _____

8. 75,391 _____

Saxon Math Intermediate 4

Name _____

Reteaching 4

Lesson 4

- # Place Value

  - Separate a three-digit number, such as money amount, into hundreds, tens, and ones.

    Example:

    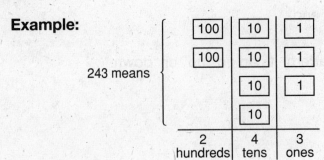

  - We can use money manipulatives to understand place value.

Practice:

1. Show $132.

Hundreds	Tens	Ones

2. Show $324.

Hundreds	Tens	Ones

   Which is less: $132 or $324? Remember to write the dollar sign. _____

3. The digit 4 is in what place in each of these numbers?

   a. 4̲1 _____   **b.** 53̲4 _____   **c.** 4̲83 _____

4. 6 hundreds, 8 tens, and 3 ones equals _____.

Reteaching 5

Lesson 5

- **Ordinal Numbers**
- **Months of the Year**

- **Ordinal numbers** tell position or order. One common use is to name days of the month and months of the year.
- Most ordinal numbers end in "th." The ordinals circled below are exceptions.

first 1st	sixth 6th	eleventh 11th
second 2nd	seventh 7th	twelfth 12th
third 3rd	eighth 8th	thirteenth 13th
fourth 4th	ninth 9th	twentieth 20th
fifth 5th	tenth 10th	twenty-first 21st

- There are 12 months in a year.
- The month/day/year form of June 12, 1998, is 6/12/98.
- We can use ordinal numbers to name the months in order. January is the first month. December is the twelfth month.

Practice:

1. Identify which circle the arrow is pointing to.

   a. ○ ○ ○ ○ ○ ○
      1st ↑

      The _____ circle.

   b. ○ ○ ○ ○ ○ ○ ○ ○
      1st   ↑

      The _____ circle.

2. On what date where you born?

   _____/_____/_____
   Month Day Year

3. This year, Martin Luther King Day is

   _____/_____/_____
   Month Day Year

4. How many months are in a year? _____

5. What is the name of the fifth month? _____

6. What is the name of the eighth month? _____

7. Which month gets the extra day? _____

8. Write the twenty-first day of September, 2007 in month/day/year form.

   _____ / _____ / _____

9. List three ordinal numbers that do not end in "th", such as "2nd".

   _____ _____ _____

Reteaching 6

Lesson 6

- **Review of Subtraction**

  - The answer to a subtraction problem is called the **difference**.

    $$\begin{array}{r} 7 \\ -\ 5 \\ \hline 2 \end{array} \text{ difference}$$

  - Check subtraction by adding.

    Subtract Down
    Seven minus five equals two.

    $$\begin{array}{r} 7 \\ -\ 5 \\ \hline 2 \end{array}$$

    Add Up
    Two plus five equals seven.

  - The **order** of numbers in subtraction is important.

    $7 - 5$ is different from $5 - 7$.

  - When you learn one **fact family**, you know four facts.

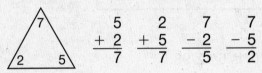

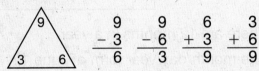

Practice:

Subtract. Check your answers by adding.

1. $\begin{array}{r} 17 \\ -\ 8 \\ \hline \end{array}$ Check: $\begin{array}{r} 8 \\ +\ \\ \hline \end{array}$ 2. $\begin{array}{r} 12 \\ -\ 7 \\ \hline \end{array}$ Check: $\begin{array}{r} 7 \\ +\ \\ \hline \end{array}$

3. $\begin{array}{r} 14 \\ -\ 6 \\ \hline \end{array}$ Check: $\begin{array}{r} 6 \\ +\ \\ \hline \end{array}$ 4. $\begin{array}{r} 18 \\ -\ 9 \\ \hline \end{array}$ Check: $\begin{array}{r} 9 \\ +\ \\ \hline \end{array}$

5. $\begin{array}{r} 11 \\ -\ 4 \\ \hline \end{array}$ Check: $\begin{array}{r} 4 \\ +\ \\ \hline \end{array}$ 6. $\begin{array}{r} 15 \\ -\ 8 \\ \hline \end{array}$ Check: $\begin{array}{r} 8 \\ +\ \\ \hline \end{array}$

7. Describe how to check a subtraction answer. Show an example.

    You can check subtraction by _____.

    Example:

Name _____

Reteaching 7
Lesson 7

- **Writing Numbers Through 999**

  - **Whole numbers** are the counting numbers and the number zero.

    0, 1, 2, 3, 4, 5, …

0	zero	10	ten	20	twenty
1	one	11	eleven	30	thirty
2	two	12	twelve	40	forty
3	three	13	thirteen	50	fifty
4	four	14	fourteen	60	sixty
5	five	15	fifteen	70	seventy
6	six	16	sixteen	80	eighty
7	seven	17	seventeen	90	ninety
8	eight	18	eighteen	100	one hundred
9	nine	19	nineteen		

- Use hyphens when writing the numbers 21–99 (except numbers that end with 0).

    426   four hundred twenty-six
    809   eight hundred nine

- Don't write "and" unless you mean a decimal point.

    $2.78   two dollars *and* seventy-eight cents

Practice:

Use words to write each number.

1. 3 _____

2. 75 _____

3. 88 _____

4. 367 _____

5. 629 _____

Use digits to write each number.

6. fifteen ____ ____

7. thirty-seven ____ ____

8. one hundred seven ____ ____ ____

9. three hundred sixty-two ____ ____ ____

Saxon Math Intermediate 4   © Harcourt Achieve Inc. and Stephen Hake. All rights reserved.

Name _____

Reteaching 8
Lesson 8

- **Adding Money**

  - Money amounts are sometimes written as two-digit numbers when there are no coins. For example, twenty-five dollars might be written $25.
  - To add money amounts:

    1. Add the ones.
    2. Add the tens.
    3. Write the dollar sign.

    Example: Sumika had $26. Then on her birthday she was given $13. How much money does Sumika have now?

    Solution: We can use $10 bills and $1 bills to add $13 to $26.

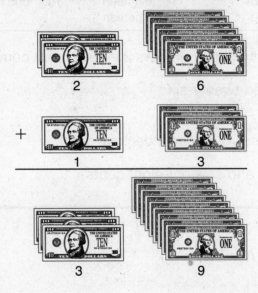

The total is 3 tens and 9 ones, which is $39.

Practice:

Add. Remember to write the dollar sign.

1. $48
   + $ 9

2. $31
   + $12

3. $72
   + $24

4. $53
   + $36

5. $27
   + $67

6. $63
   + $22

7. $51
   + $43

8. $95
   + $ 2

Name _____

Reteaching 9

Lesson 9

• Adding with Regrouping

- When added numbers in the ones column add up to more than 10 we can regroup the ones to make tens. Then we carry the new tens into the tens column.
- Regroup 10 ones to make 1 ten.

$$\begin{array}{r} 1 \\ 48 \\ +15 \\ \hline 3 \end{array}$$

1. Add ones.
   8 + 5 = 13
2. Write the 3 and carry the 1 ten to the tens column.

$$\begin{array}{r} 1 \\ 57 \\ +29 \\ \hline 6 \end{array}$$

1. Add ones.
   7 + 9 = 16
2. Write the 6 and carry the 1 ten to the tens column.

$$\begin{array}{r} 1 \\ 48 \\ +15 \\ \hline 63 \end{array}$$

3. Add tens.
   1 + 4 + 1 = 6
4. Write the 6.

$$\begin{array}{r} 1 \\ 57 \\ +29 \\ \hline 86 \end{array}$$

3. Add tens.
   1 + 5 + 2 = 8
4. Write the 8.

Practice:

Solve each problem using money manipulatives. Then add by regrouping to solve. Remember to write the dollar sign.

1. $72
   + $19

2. $38
   + $24

3. $67
   + $35

4. $42
   + $39

5. $65
   + $25

6. $51
   + $49

Use pencil and paper to add.

7. $72
   + $16

8. $75
   + $66

9. $24
   + $57

Name _____

Reteaching 10
Lesson 10

- **Even and Odd Numbers**

  - **Even numbers:** 0, 2, 4, 6, 8, ...
  - **Odd numbers:** 1, 3, 5, 7, 9, ...
  - Look at the *last* digit:

    38<u>3</u>   odd
    65<u>4</u>   even
    29<u>5</u>   odd

Practice:

Write "even" or "odd" for each number.

1. 72 _____   2. 781 _____   3. 490 _____

4. 15 _____   5. 213 _____   6. 1082 _____

7. List the five three-digit even numbers that have an 8 in the hundreds place and a 5 in the tens place.

   a. __8__ __5__ ___   b. ___ ___ ___

   c. ___ ___ ___   d. ___ ___ ___   e. ___ ___ ___

8. List the five three-digit odd numbers that have a 4 in the hundreds place and a 9 in the tens place.

   a. __4__ __9__ ___   b. ___ ___ ___

   c. ___ ___ ___   d. ___ ___ ___   e. ___ ___ ___

9. Write a three-digit even number. Write the number in words.

   ___ ___ ___

   Words _____

10. Write a three-digit odd number.

    ___ ___ ___

    Words _____

Name _____

Reteaching Inv. 1
Investigation 1

• Number Lines

- To draw a number line, begin by drawing a line. Next, put tick marks on the line, keeping an equal distance between the marks. Then label the tick marks with numbers. Sometimes every mark can be labeled and on other number lines only some marks are labeled. The mark may be labeled by two, by four, by five, or by some other number. The labels are to indicate how far the mark is from zero.

    Example: What number is the arrow pointing to?

    Begin at zero and count by ones, the distance from one tick mark to the next is 1. The arrow is pointing to the number 9.

- Sometimes zero is not shown on the number line, so we must begin counting with the number shown and find the pattern of the numbers.
- Numbers that are greater than zero are called **positive** numbers and numbers that are less than zero are called **negative** numbers. To write a negative number, we write the negative sign (minus sign) to the left of the digit.
- Zero is neither positive nor negative.

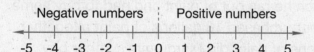

Practice:

To what number is each arrow point in problems 1–2?

1.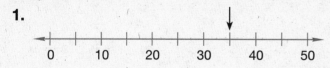

2.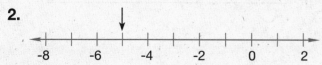

Compare:

3. 4 ◯ −5   4. −8 ◯ −4   5. 0 ◯ −7

6. Create a number line in problem 2 to arrange these numbers in order from least to greatest: 7, −5, 3, −9, 0, −1

Reteaching 11

Lesson 11

• Addition Word Problems with Missing Addends

- Addition formula: Some + Some more = Total

$$\begin{array}{r} 8 \text{ airplanes} \\ + 6 \text{ airplanes} \\ \hline 14 \text{ airplanes} \end{array} \begin{array}{l} \text{addend} \\ \text{addend} \\ \text{sum} \end{array}$$

- If either one of the **addends** is missing, we subtract the known addend from the sum.

$$\begin{array}{r} 8 \text{ airplanes} \\ + x \text{ airplanes} \\ \hline 14 \text{ airplanes} \end{array} \rightarrow \begin{array}{r} 14 \\ - 8 \\ \hline x = 6 \end{array}$$

$$\begin{array}{r} m \text{ airplanes} \\ + 6 \text{ airplanes} \\ \hline 14 \text{ airplanes} \end{array} \rightarrow \begin{array}{r} 14 \\ - 6 \\ \hline m = 8 \end{array}$$

Practice:

1. Samantha had 5 books she borrowed from the library. Her brother Brandon gave her some more books he borrowed. Samantha had a total of 12 books. How many library books did Brandon give her? _____

$$\begin{array}{r} 5 \\ + n \\ \hline 12 \end{array}$$

2. Marguerite's father bought 7 gallons of paint for the fence. After lunch he ran out of paint and had to buy some more. Her father purchased 16 gallons of paint altogether. How many gallons of paint did Marguerite's father buy after lunch? _____

$$\begin{array}{r} 7 \\ + n \\ \hline 16 \end{array}$$

Show the subtraction problem you would use to solve for the missing addend.

3. $5 + n = 13$ Subtraction problem: _____

4. $\begin{array}{r} 8 \\ + m \\ \hline 12 \end{array}$ Subtraction problem: _____

5. $w + 7 = 15$

6. $\begin{array}{r} w \\ + 4 \\ \hline 10 \end{array}$

Name _____

Reteaching 12
Lesson 12

- **Missing Numbers in Subtraction**

  - To find missing numbers in subtraction:
    When the top (first) number is missing, add.

    $$\begin{array}{r} a \\ -\ 4 \\ \hline 7 \end{array} \rightarrow \begin{array}{r} 7 \\ +\ 4 \\ \hline a = 11 \end{array}$$

    When the bottom (second) number is missing, subtract.

    $$\begin{array}{r} 15 \\ -\ n \\ \hline 9 \end{array} \rightarrow \begin{array}{r} 15 \\ -\ 9 \\ \hline n = 6 \end{array}$$

Practice:

For problems 1–4, select the operation you need to use and then find each missing number. Check your answers.

1. $\begin{array}{r} 13 \\ -\ b \\ \hline 7 \end{array}$ $b =$ ____   Check: $\begin{array}{r} 7 \\ +\ \\ \hline 13 \end{array}$

2. $\begin{array}{r} n \\ -\ 9 \\ \hline 2 \end{array}$ $n =$ ____   Check: $\begin{array}{r} 9 \\ +\ 2 \\ \hline \end{array}$

3. $\begin{array}{r} 18 \\ -\ x \\ \hline 9 \end{array}$ $x =$ ____   Check: $\begin{array}{r} 9 \\ +\ \\ \hline 18 \end{array}$

4. $\begin{array}{r} n \\ -\ 5 \\ \hline 10 \end{array}$ $n =$ ____   Check: $\begin{array}{r} 10 \\ +\ 5 \\ \hline \end{array}$

5. Write a subtraction problem with the bottom number missing. Solve and check your answer.

Find each missing number..

6. $w - 6 = 2$

7. $6 - y = 2$

8. $m - 6 = 9$

9. $9 - n = 6$

Saxon Math Intermediate 4

Name _____

Reteaching 13
Lesson 13

- **Adding Three-Digit Numbers**

  - To add three-digit numbers, add numbers in columns from right to left, starting with the ones. Regroup and carry 10s to the next column.

    Examples:
    1. Add ones.
    2. Add tens.
    3. Add hundreds.

    Show regrouping above. →
    ```
     1 1
     $675
    +$175
     $850
    ```

    1. Add ones.
    2. Add tens.
    3. Add hundreds.

    Show regrouping above. →
    ```
     1 1
     $496
    +$374
     $870
    ```

Practice:

Add. Remember to write the dollar sign in money problems.

1. $358
   + $156

2. $719
   + $208

3.   674
   + 385

4.   268
   + 392

5. $836
   + $199

6.   777
   + 232

7. $712
   + $375

8. $555
   + $445

9.   101
   + 199

14 © Harcourt Achieve Inc. and Stephen Hake. All rights reserved. Saxon Math Intermediate 4

Name _____

Reteaching 14
Lesson 14

- **Subtracting Two-Digit and Three-Digit Numbers**
- **Missing Two-Digit Addends**

Subtracting Two-Digit and Three-Digit Numbers
- To subtract three-digit numbers, work in one column at a time, starting with the ones:

  Example:
  1. Subtract ones.
  2. Subtract tens.
  3. Subtract hundreds.

  $$\begin{array}{r} 486 \\ -375 \\ \hline 111 \end{array}$$

Missing Two-Digit Addends
- To find a missing addend, always **subtract**.

  Examples:
  $$\begin{array}{r} 68 \\ +\ a \\ \hline 96 \end{array} \rightarrow \begin{array}{r} 96 \\ -68 \\ \hline a = 28 \end{array} \qquad \begin{array}{r} n \\ +32 \\ \hline 83 \end{array} \rightarrow \begin{array}{r} 83 \\ -32 \\ \hline n = 51 \end{array}$$

Practice:

Remember to write the dollar sign in money problems.

1. $\begin{array}{r} \$257 \\ -\ \$143 \end{array}$
2. $\begin{array}{r} \$678 \\ -\ \$214 \end{array}$
3. $\begin{array}{r} 576 \\ -\ 326 \end{array}$

4. $\begin{array}{r} 488 \\ -\ 223 \end{array}$
5. $\begin{array}{r} \$857 \\ -\ \$746 \end{array}$
6. $\begin{array}{r} 666 \\ -\ 444 \end{array}$

7. $\begin{array}{r} m \\ +\ 31 \\ \hline 48 \end{array} \qquad \begin{array}{r} 48 \\ -\ 31 \end{array}$

    $m =$ _____

8. $\begin{array}{r} 45 \\ +\ x \\ \hline 78 \end{array} \qquad \begin{array}{r} 78 \\ -\ 45 \end{array}$

    $x =$ _____

Name _____

Reteaching 1
Lesson 15

- ## Subtracting Two-Digit Numbers with Regrouping

  - When the top number in the ones column is less than the bottom number, we regroup by taking one 10 and moving it to the ones column.

    Examples:

    $$\begin{array}{r} \overset{6\ \ 15}{6\cancel{7}\cancel{5}} \\ -1\ 5\ 7 \\ \hline 5\ 1\ 8 \end{array} \qquad \begin{array}{r} \overset{7\ \ 13}{7\cancel{8}\cancel{3}} \\ -4\ 7\ 8 \\ \hline 3\ 0\ 5 \end{array}$$

Practice:

Use money manipulatives to model each subtraction. Then solve on paper. Remember to write the dollar sign in money problems.

1. $582
    − $456

2. $664
    − $247

3. 571
    − 364

4. 280
    − 123

5. $855
    − $746

6. 666
    − 447

7. 590
    − 382

8. 697
    − 258

9. 126
    − 99

Name _____

Reteaching 16
Lesson 16

- **Expanded Form**
- **More on Missing Numbers in Subtraction**

Expanded Form
- To express a number in **expanded form** we separate it into place values. The number 255 means "2 hundreds plus 5 tens plus 5 ones".
  We can write this in expanded form as: 200 + 50 + 5.

  Example: Write 368 in expanded form.
  300 + 60 + 8

  Example: Write 603 in expanded form. There are zero tens.
  600 + 3

Missing Numbers in Subtraction
- To find missing numbers in subtraction:

  If the top (first) number is missing, add.
  If the bottom (second) number is missing, subtract.

  Example:
  $$\begin{array}{r} a \\ -5 \\ \hline 13 \end{array} \rightarrow \begin{array}{r} 13 \\ +5 \\ \hline a = 18 \end{array}$$

Practice:

Write each number in expanded form.

1. 764 _____ + _____ + _____

2. 519 _____ + _____ + _____

3. 406 _____ + _____

4. 610 _____ + _____

Find the missing number in the subtraction problem.

5. $\begin{array}{r} 26 \\ -w \\ \hline 15 \end{array}$ $\begin{array}{r} \\ - \\ \hline \end{array}$

   $w =$ _____

6. $\begin{array}{r} p \\ -26 \\ \hline 15 \end{array}$ $\begin{array}{r} \\ + \\ \hline \end{array}$

   $p =$ _____

7. $n - 25 = 64$

   $n =$ _____

8. $45 - x = 28$

   $x =$ _____

Name _____

Reteaching Lesson 17

- **Adding Columns of Numbers with Regrouping**

- Regroup from the ones to the tens column.

```
Regroup →  4      Finding sets of 10 will help.
           45
           58
           17
           23
           64
           39
         + 86
          332
```

Practice:

Add.

1. 65
    47
    19
    + 28

2. 73
    29
    46
    + 11

3. 42
    68
    27
    54
    + 86

4. 96
    54
    32
    17
    + 24

5. 32
    67
    49
    23
    + 81

6. 41
    58
    15
    76
    + 39

Name _____

Reteaching 18
Lesson 18

- **Temperature**

  - A **scale** is a type of number line often used for measuring. Scales are found on rulers, gauges, thermometers, speedometers, and many other instruments.
  - We use a thermometer to measure **temperature**. Temperature is usually measured in **degrees Fahrenheit (°F)** or in **degrees Celsius (°C)**.
  - To read the temperature on a thermometer, try different skip counts to find the interval. On a thermometer, **tick marks** are often two degrees apart.

    Example: What temperature is shown by this thermometer?

    First, find the interval. Counting by 2s matches the marking on the scale.

    Count up by 2s. The temperature is 42°F.

Practice:

What measurement is shown on each of these scales? Remember to write the units.

1.  _____

2.   _____

3. Jeremy reads the thermometer at 8:00 a.m. and records a temperature of 68°F. At 9:00 a.m., the temperature is 14° warmer. Shade in the thermometer to show the temperature at 9:00 a.m.

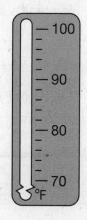

Name _____

Reteaching 19
Lesson 19

- **Elapsed-Time Problems**

  - The **short hand** tells the **hour**.
  - The **long hand** tells the **minutes**.
  - Count by 5s to find the number of minutes as the long hand moves from one whole number to the next.

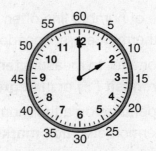

  - A "quarter" in **time** is 15 minutes because 15 minutes is one quarter $(\frac{1}{4})$ of an hour.

"Half past seven"

"A quarter **after** six"

"A quarter **to** four"

  - **a.m.** = 12 hours before noon
  - **p.m.** = 12 hours after noon

Practice:

If it is morning, what time is shown by each clock? Remember to write "a.m." or "p.m."

1. _____

2. _____

3. _____

4. Use digital form to show what time it is at "ten minutes to seven" in the evening.

   _____

5. Use digital form to show what time it is at "twenty-five minutes after three" in the afternoon.

   _____

Name _____

Reteaching 20
Lesson 20

- **Rounding**

  - To round a number to the nearest ten:
      1. Look at the ones place.
      2. Ask: Is it 5 or more? (5, 6, 7, 8, 9)
          Yes → Add 1 to the tens.
          No → The tens place stays the same.
      3. Replace the digit in the ones place with zero.

      Examples: 6<u>3</u> → 60   6<u>6</u> → 70

  - To round dollars and cents to the nearest whole dollar:
      1. Look at cents. Is it 50¢ or more?
          Yes → Add one to the dollar amount.
          No → Keep the dollar amount the same.
      2. In both cases, drop the cents and the decimal.

      Examples: $6.85 → $7 $6.42 → $6

Practice:

Round each number to the nearest ten.

1. 8<u>3</u> → _____   2. 2<u>9</u> → _____   3. 5<u>5</u> → _____

Round each amount of money to the nearest dollar. Remember to write the dollar sign in money problems.

4. $8.73 → _____   5. $6.28 → _____   6. $3.54 → _____

7. $10.84 → _____   8. $18.46 → _____   9. $29.96 → _____

Saxon Math Intermediate 4

Name _____

Reteaching Inv

Investigation 2

Units of Length and Perimeter

- Two systems of units are used to measure length:

U.S. Customary
Some of the units in this system are inches, feet, yards, and miles.

Metric
Some of the units in this system are millimeters, centimeters, meters, and kilometers.

Example: This line segment measures 25 mm on a metric ruler and about 1 inch on a customary ruler.

- Perimeter is the distance around a shape.
- Add all sides $P = l + w + l + w$

Practice:

1. Using your ruler, how many inches long and wide is this rectangle below?

2. What is the perimeter of this rectangle?

3. What is the perimeter of this rectangle?

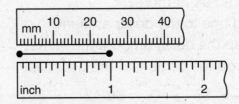

4. What is the perimeter of a square that has sides 5 inches long?

Name _____

Reteaching 21

Lesson 21

- **Triangles, Rectangles, Squares, and Circles**

  - **Triangle** → 3 sides   **Rectangle** → 4 sides

                                               width
                                            length

  Square → a rectangle with 4 **equal** sides

- Measures of a **circle**:

  Radius = $\frac{1}{2}$ × diameter   **Diameter** = 2 × radius

  Radius   Diameter

- To draw circles, we can use a tool called a **compass**.
  Here are two types of compasses:

Practice:

1. Draw a triangle with no sides that are the same length.

2. Draw a rectangle that is about three times as long as it is wide.

3. Use a compass to draw a circle with a diameter of 2 inches.

4. Draw a square that has sides 2 inches long.

Name _____

Reteaching 22
Lesson 22

- **Naming Fractions**
- **Adding Dollars and Cents**

Naming Fractions
- To find the fraction of a shape that is shaded:
    1. Count the number of shaded parts. → top number
    2. Count the total number of parts. → bottom number

Example:

$$\frac{\text{parts shaded}}{\text{total number of parts}} \quad \frac{3}{4} \quad \frac{\textbf{numerator (top number)}}{\textbf{denominator (bottom number)}}$$

 $\frac{1}{3}$ one third $\frac{1}{4}$ one fourth $\frac{1}{10}$ one tenth

 $\frac{3}{5}$ three fifths $\frac{5}{6}$ five sixths $\frac{7}{8}$ seven eighths

Adding Dollars and Cents
- To add dollars and cents, start with pennies.

Example: 1. Add pennies.
2. Add dimes.
3. Add dollars.

```
   1 1
   $3.56   Line up the
 + $3.54   decimal points.
   ─────
   $7.10
```

- Remember to write the dollar sign and decimal point in the sum.

Practice:

What fraction of each shape is shaded?

1. _____   2. _____   3. 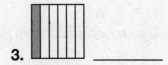 _____

4. _____   5. _____   6. _____

7.    8.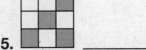

Reteaching 23

Lesson 23

- **Lines, Segments, Rays, and Angles**

  - A **line** extends in opposite directions with no end. Arrowheads show that it continues in both directions.

    line

  - A line **segment** is part of a line. It has **endpoints**, not arrowheads.

    segment

  - A **ray** begins at a point and continues in one direction without end. It has one arrowhead.

    ray

  - **Parallel** lines or segments never cross.
  - When lines or segments cross, we say they **intersect**.
  - Intersecting lines or segments that form "square corners" are **perpendicular**.

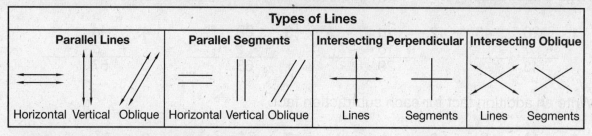

  - **Angles** are formed where lines or segments intersect or where two or more rays or segments begin.

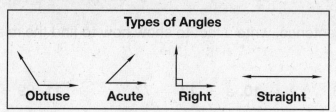

Practice:

1. Draw two segments that intersect and are perpendicular.

2. Draw a ray.

3. Describe something in the real world that can represent a pair of parallel lines.

Name _____

Reteaching 24

Lesson 24

- **Inverse Operations**

  - When we know one addition fact, we know three other facts.

    Example: If we know $n + 1 = 3$

    then we also know $1 + n = 3$ $3 - n = 1$ $3 - 1 = n$

  - Notice that one of the facts shows us how to find the missing addend from the original problem.

    $3 - 1 = n$ So, $n = 2$

  - Addition and subtraction are **inverse operations** because one operation "undoes" or "reverses" the other.

Practice:

Write a subtraction fact for each addition fact.

1. $\begin{array}{r} 26 \\ +\ r \\ \hline 43 \end{array}$ $\begin{array}{r} 43 \\ -\ \_\_ \\ \hline \_\_ \end{array}$

2. $\begin{array}{r} m \\ +\ 15 \\ \hline 29 \end{array}$ $\begin{array}{r} \_\_ \\ -\ \_\_ \\ \hline \_\_ \end{array}$

3. $\begin{array}{r} 39 \\ +\ z \\ \hline 63 \end{array}$ $\begin{array}{r} \_\_ \\ -\ \_\_ \\ \hline \_\_ \end{array}$

4. $\begin{array}{r} 44 \\ +\ d \\ \hline 57 \end{array}$ $\begin{array}{r} \_\_ \\ -\ \_\_ \\ \hline \_\_ \end{array}$

Write an addition fact for each subtraction fact.

5. $\begin{array}{r} 43 \\ -\ q \\ \hline 6 \end{array}$ $\begin{array}{r} 6 \\ +\ \_\_ \\ \hline \_\_ \end{array}$

6. $\begin{array}{r} t \\ -\ 19 \\ \hline 38 \end{array}$ $\begin{array}{r} \_\_ \\ +\ \_\_ \\ \hline \_\_ \end{array}$

7. $\begin{array}{r} 64 \\ -\ a \\ \hline 26 \end{array}$ $\begin{array}{r} \_\_ \\ +\ \_\_ \\ \hline \_\_ \end{array}$

8. $\begin{array}{r} 17 \\ -\ w \\ \hline 3 \end{array}$ $\begin{array}{r} \_\_ \\ +\ \_\_ \\ \hline \_\_ \end{array}$

For each number sentence, write a fact to show how to find the missing number. Then solve.

9. $23 + t = 46$

    _____

    $t =$ _____

10. $a + 12 = 77$

    _____

    $a =$ _____

11. $99 - y = 9$

    _____

    $y =$ _____

Name _____

Reteaching 25
Lesson 25

- **Subtraction Word Problems**

  - Subtraction problems follow a pattern:
    "Some − Some went away = Some left"

  - Another way to express the pattern is:
    "Original amount − Some part = Difference"

  - If the original amount (top number) is missing, add the **difference** to the part.

Some	m apples	23
Some went away	− 12 apples	+ 12
Some left	23 apples	35

    $m = 35$ apples

  - If the subtracted part is missing, subtract the difference from the original amount.

Some	45 apples	45
Some went away	− m apples	− 28
Some left	28 apples	17

    $m = 17$ apples

  - If the difference is missing, subtract the part from the original amount.

Some	67 apples	67
Some went away	− 34 apples	− 34
Some left	m apples	33

    $m = 33$ apples

Practice:

1. At the start line, 53 cyclists had water. Some cyclists dropped their bottles during the race. At the finish, only 28 cyclists had bottles. How many cyclists dropped bottles?

53	had bottles
− w	dropped bottles
28	now have bottles

    $w =$ _____ dropped bottles

2. A flock of geese started flying north. Then 55 geese landed at a pond. Now 28 geese are flying together. How many geese were flying north before some landed?

y	geese started
− 55	landed
28	now flying

    $y =$ _____ geese started

3. Thom had $40. He spent $24. Then how much money did Thom have?

Saxon Math Intermediate 4

Reteaching Lesson 26

- **Drawing Pictures of Fractions**

  - To draw a picture of a fraction:
    1. Draw the figure.
    2. Divide into **equal parts**.
    3. Shade the correct number of parts.

  Examples:

  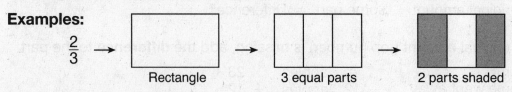

  Other examples:

  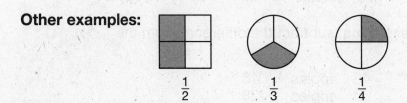

  - To divide a circle into equal **thirds**:
    1. Draw a dot in the center.   These are not equal parts:
    2. Make a "Y" from the dot.

Practice:

1. Shade one fourth of the square.

2. Shade two thirds of the circle.

3. Shade two fifths of the rectangle.

4. Shade three fourths of the circle.

5. Is one fifth of this circle shaded? _____

   Why or why not? _____

   _____

Name _____

Reteaching 27
Lesson 27

- **Multiplication as Repeated Addition**
- **More Elapsed-Time Problems**

Multiplication as Repeated Addition
- Multiplication can represent the addition of identical numbers.

    Example: $5 + 5 + 5 + 5 + 5 + 5 = 6 \times 5$

Elapsed Time
- Picture a clock face divided into 4 equal parts. Each part represents 15 minutes.
- The difference between a time on the clock and the time directly across from it is always 30 minutes.
- Every **twelve hours** is the same hour—only the a.m. or p.m. will change.
- Every **twenty-four hours** is the same time of day, but it is the next day of the week.

Practice:

Write the following repeated addition problems as multiplication problems.

1. $2 + 2 + 2 + 2 + 2 + 2 + 2 =$ _____

2. $8 + 8 + 8 + 8 =$ _____

Write the following multiplication problems as repeated addition problems.

3. 6×3 _____   4. 4×5 _____   5. 3×8 _____

Look at a clock or use a student clock to answer problems 6–7. Remember to write "a.m." or " p.m."

6. If it is morning, what time will it be in 3 hours and 45 minutes?

    Start time: 10:45

    Count forward 45 minutes. _____

    Count forward 3 hours. _____

7. If it is evening, what time was it 7 hours and 15 minutes ago?

    Start time: 8:15

    Count backward 15 minutes. _____

    Count backward 7 hours. _____

Name _____

Reteaching 28
Lesson 28

- # Multiplication Table

  - Numbers we multiply together are called **factors**. The answer to a multiplication problem is called a **product**.
  - A multiplication table shows the products of different pairs of factors.
  - To use the multiplication table to find a product, we first find one factor in a row. Then we find the other in a column. The product is the number where the row and column meet.
  - The **Commutative Property of Multiplication** states that changing the order of factors does not change the product.
    $a \times b = b \times a$
  - The **Property of Zero for Multiplication** states that any number times zero equals zero.
    $a \times 0 = 0$
  - The **Identity Property of Multiplication** states that any number times one equals the number.
    $a \times 1 = a$

Properties of Multiplication

Commutative Property	$m \times n = n \times m$
Identity Property	$1 \times n = n$
Zero Property	$0 \times n = 0$

Practice:

Use the multiplication table to find each product.

1. 9
    × 6

2. 6
    × 9

3. 8
    × 4

4. 4
    × 8

5. 7
    × 8

6. 8
    × 7

7. 5
    × 9

8. 9
    × 5

Name _____

Reteaching 29
Lesson 29

- **Multiplication Facts (0s, 1s, 2s, 5s)**

- Zero × any number = 0. (Property of Zero for Multiplication)
- One × any number = the same number. (Identity Property of Multiplication)
- Two × any number = double the number.
- Five × any number = a number that ends in 0 or in 5.

Practice:

Complete the multiplication facts below.

1. 5 × 3 = _____
2. 8 × 2 = _____
3. 0 × 4 = _____

4. 1 × 7 = _____
5. 7 × 5 = _____
6. 3 × 2 = _____

7. 1 × 5 = _____
8. 4 × 1 = _____
9. 6 × 2 = _____

10. 5 × 4 = _____
11. 9 × 5 = _____
12. 2 × 2 = _____

13. 6 × 1 = _____
14. 3 × 4 = _____
15. 5 × 0 = _____

16. 0 × 7 = _____
17. 5 × 6 = _____
18. 2 × 7 = _____

19. 5 × 5 = _____
20. 1 × 2 = _____
21. 2 × 5 = _____

Reteaching 30

Lesson 30

- **Subtracting Three-Digit Numbers with Regrouping**

  - Work from right to left.
  - When the digit in the top number is smaller than the digit in the bottom number, we regroup from the next place to the left.
  - When regrouping, it helps to cross-out the digit and rewrite the new number above the column.
  - When subtracting dollars and cents, remember to line up the decimal points and to write the dollar sign in money problems.

Example:

$$\begin{array}{r}{}^{0\ 17}\\ \$5.\cancel{1}\cancel{7}\\ -\$3.28\\ \hline 9\end{array} \rightarrow \begin{array}{r}{}^{4\ 10\ 17}\\ \$\cancel{5}.\cancel{1}\cancel{7}\\ -\$3.28\\ \hline .89\end{array} \rightarrow \begin{array}{r}{}^{4\ 10\ 17}\\ \$\cancel{5}.\cancel{1}\cancel{7}\\ -\$3.28\\ \hline \$1.89\end{array}$$

$$\rightarrow \begin{array}{r}{}^{3\ 13}\\ \$6.4\cancel{3}\\ -\$4.56\\ \hline 7\end{array} \rightarrow \begin{array}{r}{}^{5\ 13\ 13}\\ \$\cancel{6}.\cancel{4}\cancel{3}\\ -\$4.56\\ \hline .87\end{array} \rightarrow \begin{array}{r}{}^{5\ 13\ 13}\\ \$\cancel{6}.\cancel{4}\cancel{3}\\ -\$4.56\\ \hline \$1.87\end{array}$$

Practice:

Subtract. Remember to write the dollar sign in money problems.

1. $451
    − $277

2. $6.74
    − $4.75

3. 583
    − 396

4. 340
    − 157

5. 449
    − 299

6. 982
    − 695

Name _____

Reteaching Inv. 3
Investigation 3

- **Area Models**

  - An array is a rectangular arrangement of numbers or symbols in columns and rows.

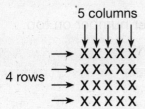

This array shows that 4 times 5 equals 20. It also shows that 4 and 5 are both factors of 20.

  - An area model is an array of connected squares.

    This model shows that 6 × 3 = 18.

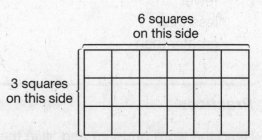

  - **Area** of a shape is measured by counting the number of squares of a certain size that are needed to cover a surface.

  - A square is a rectangle whose length and width are equal. We say that we "square a number" when we multiply by a number by itself.

  - The numbers 1, 4, 9, 16, 25, 36, and so on form a sequence of **square numbers,** or **perfect squares.**

  - To find a **square root** of a number, we find a number that, when multiplied by itself, equals the original number. The square root of a number is indicated using the square root symbol $\sqrt{}$.

    For example, "the square root of twenty-five equals five" or $\sqrt{25} = 5$.

Practice:

1. Draw a 6 column by 3 row array of Xs. What multiplication fact is represented by the array?

2. Draw an array of 15 Xs arranged in 5 rows. Then write the multiplication fact.

3. Draw an area model of 3 × 4. How many small squares are in the rectangle?

4. What is the multiplication fact that is illustrated by this rectangle?

Find each square root.

5. $\sqrt{36}$ 6. $\sqrt{81}$ 7. $\sqrt{100}$

Name _____

Reteaching 3
Lesson 31

- **Word Problems About Comparing**

    - To find the difference between two numbers, subtract.

Formula	Problem
Larger	52 apples
− Smaller	− 21 apples
Difference	31 apples

        Always put the larger number on top.

    - Watch for these words:

        more
        fewer
        less than
        greater than

Practice:

Read the word problem and fill in the blanks to solve the problems.

1. Cray Lake is 74 feet deep. Silver Lake is 68 feet deep. How many more feet deep is Cray Lake?

    _____ Cray Lake depth

    − _____ Silver Lake depth

    _____ feet deeper

2. Spencer had a collection of 63 comic books. His friend Annabeth had a collection of 78 comics. How many more comics does Annabeth have than Spencer?

    _____ _____

    − _____ _____

    _____ more comics

3. Paulo lives 12 minutes from the pool, while Isaac lives 26 minutes from the pool. How many minutes more does it take Isaac to get to the pool than Paulo?

    _____ _____

    − _____ _____

    _____ minutes more

© Harcourt Achieve Inc. and Stephen Hake. All rights reserved.

Saxon Math *Intermediate 4*

Name _____

Reteaching 32
Lesson 32

- **Multiplication Facts (9s, 10s, 11s, and 12s)**

  - 9s multiplication facts:

    The **first digit** of the product is **one less** than the factor.
    The **two digits** of the product always add up to **9**.

    Sets of 9
8 + 1	= 9
7 + 2	= 9
6 + 3	= 9
5 + 4	= 9

  - 10s multiplication facts:

    To multiply a whole number by 10, copy the number, then attach a **zero**.
    The **last digit** of the product is always zero.
    We can see a pattern in the multiples of 10 using the following money model:

Dimes	1	2	3	4	5	6	7	8	9	10
Pennies	10	20	30	40	50	60	70	80	90	100

  - 11s multiplication facts:

    For the numbers 1–9, both digits in the product are the factor that is not 11.
    We can see a pattern in multiples of 11 by looking at the number of players in a soccer team:

Teams	1	2	3	4	5	6	7	8	9	10
Players	11	22	33	44	55	66	77	88	99	110

    When we multiply 10 × 11, we are using the multiplication rules for 10 *and* 11.

  - 12s multiplication facts:

    The digits in the product add up to a multiple of 3. Count up by 3s to check your work.
    We can see a pattern in the multiples of 12 by thinking about feet and inches:

Feet	1	2	3	4	5	6	7	8	9	10
Inches	12	24	36	48	60	72	84	96	108	120

Practice:

Find the product for each multiplication fact.

1. 9 × 7
2. 10 × 9
3. 9 × 2
4. 12 × 5

5. 11 × 4
6. 6 × 9
7. 12 × 3
8. 11 × 9

Saxon Math Intermediate 4 © Harcourt Achieve Inc. and Stephen Hake. All rights reserved.

Name _____ **Reteaching 33**
Lesson 33

- **Writing Numbers Through Hundred Thousands**

  - Use hyphens to spell out the numbers 21–99 (except numbers that end with zero).
  - Remember to place a comma after the word **thousand**:

4507	written in words is	four thousand, five hundred seven
34,507	written in words is	thirty-four thousand, five hundred seven
234,507	written in words is	two hundred thirty-four thousand, five hundred seven

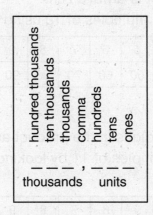

  - Counting from the right, place a comma every three digits.

    Example: 654321 → 654,321

  - Picture the numbers in groups of three.

    Each group is a family (thousands family, units family, etc.)
    Each family has three members: ones, tens, and hundreds.

Practice:

Read the following numbers to your teacher:

1. (218),000
2. (516),000
3. (16),(300)
4. (675),000
5. (6)(385)
6. (450),(295)

Use words to write these numbers:

7. (3)(112) _____

8. (21),(283) _____

9. (618),(493) _____

Name _____

Reteaching 34
Lesson 34

- **Writing Numbers Through Hundred Millions**

  - Use a comma to separate families. Three places follow every comma.

```
hundred millions
ten millions
millions
comma
hundred thousands
ten thousands
thousands
comma
hundreds
tens
ones

___ , ___ , ___
millions thousands units
```

Practice:

Use digits to write each number.

1. two million, two hundred fifteen thousand, six hundred eighty-four

   ___ , ___ ___ ___ , ___ ___ ___

2. eighteen million, forty-five thousand, eight hundred five

   ___ ___ , ___ ___ ___ , ___ ___ ___

3. four hundred one million, nine hundred seventy-six thousand, nine

   ___ ___ ___ , ___ ___ ___ , ___ ___ ___

4. eight million, seven hundred eleven, two hundred fifty-six

   ___ , ___ ___ ___ , ___ ___ ___

5. five hundred sixty-three million, eight

   ___ ___ ___ , ___ ___ ___ , ___ ___ ___

Saxon Math Intermediate 4 © Harcourt Achieve Inc. and Stephen Hake. All rights reserved.

Name _____

Reteaching 35
Lesson 35

• **Naming Mixed Numbers and Money**

Naming Mixed Numbers
• A **mixed number** is a whole number and a fraction: $3\frac{1}{2}$
• Use the word "and" when naming mixed numbers.

   $= 2\frac{1}{4}$ two *and* one fourth

• In amounts of money, we use the word "and" to read the decimal point.
The decimal point must be two digits from the end of the number.
If there are no dollars, write a zero in front of the decimal point.

$0.04 means the same as 4¢, which we read as "four cents."
40¢ = $0.40, which we read as "forty cents."
We read $1.04 as "one dollar and four cents."

Practice:

What mixed numbers are pictured here?

1.   = _____

2.   = _____

Use words to write each mixed number.

3. $10\frac{3}{4}$ _____

4. $3\frac{75}{100}$ _____

Write each amount with a **cent sign** instead of a dollar sign.

5. $0.34 _____   6. $0.71 _____

Write each amount with a **dollar sign** instead of a cent sign.

7. 63¢ _____   8. 5¢ _____

Name _____

Reteaching 36
Lesson 36

• Fractions of a Dollar

- Coins represent **fractions** of a dollar.
- One penny is one hundredth of a dollar.
    100 pennies = 1 dollar, so 3 pennies is $\frac{3}{100}$ of a dollar.
    As a dollar amount, it looks like this: $0.03

- One fourth is also called "one quarter."
    4 quarters = 1 dollar, so 1 quarter is $\frac{1}{4}$ of a dollar.
    As a dollar amount, it looks like this: $0.25

- One dime is one tenth of a dollar.
    10 dimes = 1 dollar, so 3 dimes is $\frac{3}{10}$ of a dollar.
    As a dollar amount, it looks like this: $0.30

- One nickel is one twentieth of a dollar.
    20 nickels = 1 dollar, so 7 nickels is $\frac{7}{20}$ of a dollar.
    As a dollar amount, it looks like this: $0.35

Practice:

1. Write the value of two quarters and a dime using a dollar sign and a decimal point. _____

2. Write two quarters and a dime as a fraction of a dollar. _____

3. What fraction of a dollar is three dimes? _____

4. Write the value of three nickels using a dollar sign and a decimal point. _____

5. Forty-five pennies are what fraction of a dollar? _____

6. Write the value of 45 pennies using a dollar sign and a decimal point. _____

7. Compare: 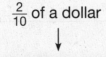 of a dollar   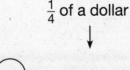 of a dollar

   _____ ¢   _____ ¢

Name _____ **Reteaching 3**
Lesson 37

- **Reading Fractions and Mixed Numbers from a Number Line**

- To name mixed numbers on a number line:
    1. Count segments from the whole number to the point to find the **numerator** (top number).
    2. Count the segments between whole numbers to find the **denominator** (bottom number).
    3. Remember to name the whole number.

Example:

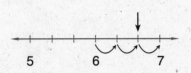

1. Two segments between 6 and the arrow. The numerator is 2.
2. Three segments between 6 and 7. The denominator is 3.
3. The whole number is 6. The answer is $6\frac{2}{3}$.

Practice:

Name each fraction or mixed number marked by the arrows below.

1. _____
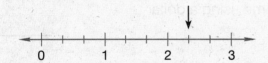

2. _____

3. _____
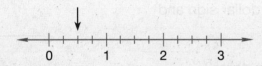

4. _____

5. _____
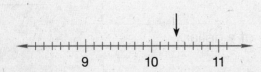

6. _____
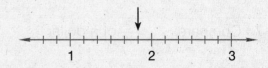

Name _____

Reteaching 38
Lesson 38

- **Multiplication Facts (Memory Group)**

    - Read and practice recalling these multiplication facts.

        3 × 4 = 12   4 × 7 = 28

        3 × 6 = 18   4 × 8 = 32

        3 × 7 = 21   6 × 7 = 42

        3 × 8 = 24   6 × 8 = 48

        4 × 6 = 24   7 × 8 = 56

- Because multiplication and division are **inverse operations**, we can find division facts to complete a fact family for each fact in the memory group.

Practice:

Write two multiplication facts and two division facts for each fact family below.

1. △(18, 6, 3)   3 × 6 = 18   ___ × ___ = ___   3)‾18   ___)‾___

2. △(32, 4, 8)   ___ × ___ = ___   ___ × ___ = ___   ___)‾___   ___)‾___

3. △(56, 8, 7)   ___ × ___ = ___   ___ × ___ = ___   ___)‾___   ___)‾___

4. △(48, 8, 6)   ___ × ___ = ___   ___ × ___ = ___   ___)‾___   ___)‾___

Multiply. Try to write the answers quickly without stopping between problems.

5. 4 × 3 _____   6. 4 × 6 _____   7. 4 × 7 _____

8. 3 × 6 _____   9. 3 × 7 _____   10. 3 × 8 _____

11. 6 × 4 _____   12. 6 × 5 _____   13. 6 × 7 _____

14. 7 × 8 _____   15. 8 × 7 _____   16. 8 × 4 _____

Name _____

Reteaching 39

Lesson 39

- **Reading an Inch Scale to the Nearest Fourth**

  - Practice counting across the ruler below.

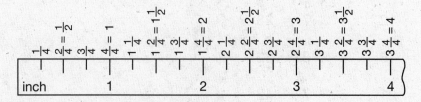

- Each **inch** on this ruler is divided into **fourths**.
- The long marks are inch marks.
- The short marks are $\frac{1}{4}$-inch marks.
- The marks halfway between the inch marks are $\frac{1}{2}$-inch marks.
- Remember, $\frac{2}{4} = \frac{1}{2}$.

Practice:

Name each point marked by an arrow on this inch scale. Remember to write the units.

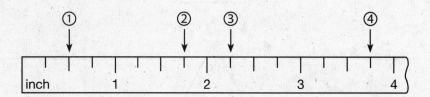

1. _____ 2. _____ 3. _____ 4. _____

Read each point marked by an arrow on this inch scale to the nearest fourth.

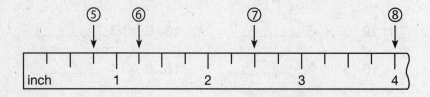

5. _____ 6. _____ 7. _____ 8. _____

Name _____

Reteaching 40
Lesson 40

• Capacity

Capacity Units		
U.S.		**Metric**
oz	ounce	mL milliliter
c	cup	L liter
pt	pint	
qt	quart	
gal	gallon	

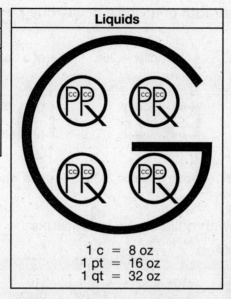

1 c = 8 oz
1 pt = 16 oz
1 qt = 32 oz

Capacity Equivalences	
U.S. Customary	**Metric**
16 oz = 1 pt	1000 mL = 1 L
2 pt = 1 qt	
4 qt = 1 gal	

1 mL   1 oz   1 cup   1 pint   1 quart   1 liter $\frac{1}{2}$ gallon   2 liters   1 gallon

---

Practice:

1. How many cups are equal to one quart? _____

2. How many pints are equal to one gallon? _____

3. How many ounces are equal to one pint? _____

4. Which is larger one quart or one liter? _____

5. How many milliliters in one liter? _____

6. How many milliliters will you need for 5 liters? _____

7. U.S. Liquid Measure

   8 fl oz = 1 c

   _____ c = 1 pt

   _____ pt = 1 qt

   _____ qt = 1 gal

Name _____

Reteaching Inv.
Investigation 4

- **Relating Tenths and Hundredths and Fractions to Decimals**

  - Just as one cent is one hundredth of a whole dollar, one percent is one hundredth of a whole.

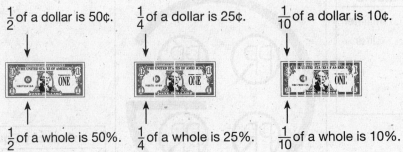

$\frac{1}{2}$ of a dollar is 50¢. $\frac{1}{4}$ of a dollar is 25¢. $\frac{1}{10}$ of a dollar is 10¢.

$\frac{1}{2}$ of a whole is 50%. $\frac{1}{4}$ of a whole is 25%. $\frac{1}{10}$ of a whole is 10%.

  - We represent these places with dollars, dimes and pennies as shown.

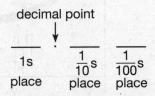

decimal point

1s place $\frac{1}{10}$s place $\frac{1}{100}$s place

  - Fractions and decimals are two ways to describe parts of a whole.
  - Fractions always show a numerator and a denominator. Decimal numbers do not show the denominator, but it is indicated by the number of places to the right of the decimal point.

    $0.1 = \frac{1}{10}$ one decimal place $0.12 = \frac{12}{100}$ two decimal places

  - Decimal numbers name the numerator shown by the digits and the denominator is named by the number of decimal places.

    $\frac{75}{100}$ is read "seventy-five hundredths"

Practice:

Use bills and coins to represent problems 1–2.

1. $13.43
2. Six and four hundredths.

Use words to name each of these numbers.

3. $\frac{63}{100}$ 4. 0.63   5. 4.8   6. 15.93

Use digits to write each of these decimal numbers.

7. six and seven tenths
8. eight and twenty-nine hundredths

Name _____

Reteaching 41
Lesson 41

- **Subtracting Across Zeros**
- **Missing Factors**

Subtracting Across Zeros
- Instead of regrouping one step at a time, we can look at the problem another way.
- Mentally group the hundreds and tens digits as tens. Think: 3 hundreds is 30 tens.
- Then regroup "one" from the "tens."
- **Regroup** across all zeros in one step:
- Compare:

Regroup one step at a time:

$$\begin{array}{r} \overset{210}{\$\cancel{3}\cancel{1}6} \\ -\ \$2\ 7\ 8 \\ \hline \$\ \ \ 3\ 8 \end{array}$$

Regroup across all places in one step:

$$\begin{array}{r} \overset{30\ \ 16}{\$\cancel{3}\cancel{1}6} \\ -\ \$2\ 7\ 8 \\ \hline \$\ \ \ 3\ 8 \end{array}$$

Missing Factors
- **Factors** are numbers that are multiplied. The **product** is the answer to a multiplication problem.

    factor × factor = product

- To find a missing factor, divide. $7n = 28$ $(7\overline{)28} = 4)$
- Letters that take the place of numbers are called **variables**.
- When a number and variable are written side by side, it means the number and variable are multiplied. So $3n$ means 3 times n.

    $3n = 15$ $8x = 48$
    $n = 5$ $x = 6$

Practice:

Subtract. Remember to write the dollar sign in money problems.

1. $\ \ \ \$721$
   $-\ \$546$

2. $\ \ \ \$2.06$
   $-\ \$1.67$

3. $\ \ \ 534$
   $-\ 355$

Find the missing factor in each problem.

4. $5m = 45$

   $m =$ _____

5. $8x = 24$

   $x =$ _____

6. $7w = 49$

   $w =$ _____

7. $3y = 36$

   $y =$ _____

Name _____

Reteaching
Lesson 42

- **Rounding Numbers to Estimate**

  - To round a number to the nearest hundred:
    1. Look at the tens place.
    2. Ask: Is the digit in the tens place 5 or more? (5, 6, 7, 8, 9)

       Yes → Add 1 to the hundreds place.
       No → The hundreds place stays the same.

    3. Replace the numbers after the hundreds place with zeros.

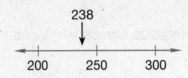

   2<u>3</u>8   3 is less than 5

   238 is closer to 200 than to 300.

   238 rounds to 200.

Practice:

Round each number to the nearest hundred.

1. <u>6</u>14 → _____
2. <u>5</u>83 → _____
3. <u>1</u>49 → _____
4. 1<u>7</u>35 → _____
5. 4<u>0</u>87 → _____
6. 50 → _____
7. 904 → _____
8. 317 → _____
9. 851 → _____

Name _____

Reteaching 43

Lesson 43

- **Adding and Subtracting Decimal Numbers, Part 1**

  - Line up the decimal points carefully.

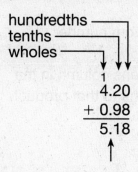
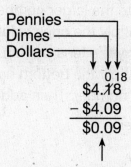

All decimal points are in line.   All decimal points are in line.

Practice:

Find each sum or difference. Remember to write the dollar sign in money problems.

1. $5.74
   + $2.38

2. $2.18
   + $1.99

3. $0.51
   + $0.60

4. 3.46
   + 3.35

5. 7.08
   − 5.49

6. 4.68
   − 2.81

7. 4.30
   − 1.26

8. 0.47
   − 0.28

9. 24.1
   − 16.5

Saxon Math Intermediate 4

Reteaching
Lesson 44

- **Multiplying Two-Digit Numbers, Part 1**

  - To multiply a two-digit number by a one-digit number:
    1. Write the larger number on top.
    2. Multiply the ones column by the bottom number.
    3. Carry the tens portion.
    4. Multiply the bottom number by the tens column in the top number. Then add the carried tens to that product.

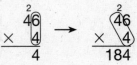

Example:

   1. 3 × 5 = 15. Write the 5.
        2. Carry the 1 (ten).

   3. 3 × 2 = 6; 6 + 1 = 7.
        Write the 7.

Practice:

Find each product.

1. 32
   × 3

2. 34
   × 4

3. 43
   × 3

4. 44
   × 5

5. 51
   × 5

6. 54
   × 2

7. 62
   × 4

8. 64
   × 2

Name _____

Reteaching 45

Lesson 45

- **Parentheses and the Associative Property**
- **Naming Lines and Segments**

Parentheses and the Associative Property
- The **Associative Property of Addition** states that how the numbers are grouped does not affect the **sum**. We do the work **inside the parentheses** first.

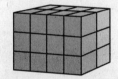   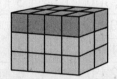
$(3 + 4) + 5 = 12$ $3 + (4 + 5) = 12$

- The **Associative Property of Multiplication** states that how the numbers are grouped does not affect the **product**.

$(3 \times 4) \times 2 = 24$ $3 \times (4 \times 2) = 24$

- The Associative Property does not apply to subtraction or division.

Naming Lines and Segments
- Name a line with two points.

$\overleftrightarrow{AB}$ or $\overleftrightarrow{BA}$

This is line *AB*. It is also line *BA*.

- Name a segment with two endpoints.

$\overline{RS}$ or $\overline{SR}$

This is segment *RS*. It is also segment *SR*.

Practice:

1. $7 - (2 + 5) =$ _____

2. $8 + (6 - 4) =$ _____

3. $5 \times (6 - 3) =$ _____

4. $3 \times (12 - 5) =$ _____

5. $6 + (5 \times 2) + 4 =$ _____

6. $4 + (3 \times 10) - 2 =$ _____

Name _____

Reteaching 4
Lesson 46

- **Relating Multiplication and Division, Part 1**

  - To find a missing factor, divide.

    Division "undoes" multiplication because division and multiplication are inverse operations.

    $5 \times w = 10$ $n \times 2 = 10$
    $w = 10 \div 5$ $n = 10 \div 2$
    $w = 2$ $n = 5$

  - A multiplication table can also be used to find missing factors.

    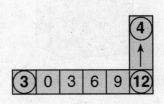

Practice:

Divide.

1. $2\overline{)10}$ 2. $6\overline{)24}$ 3. $5\overline{)30}$ 4. $7\overline{)42}$

5. $9\overline{)18}$ 6. $8\overline{)32}$ 7. $9\overline{)54}$ 8. $4\overline{)28}$

Write the three remaining facts from the fact family.

9. $4\overline{)12}$ _____ ÷ _____ = _____

                          _____ × _____ = _____

                          _____ × _____ = _____

Name _____

Reteaching 47

Lesson 47

- **Relating Multiplication and Division, Part 2**

  - With one multiplication fact we can form one more multiplication fact and two division facts.
  - Below are three ways to show eighteen **divided by** three.

    $3\overline{)18} \qquad 18 \div 3 \qquad \dfrac{18}{3}$

  - Always say the greater number (**dividend**) first.

Practice:

Divide.

1. $64 \div 8 =$ _____ 2. $45 \div 9 =$ _____ 3. $28 \div 7 =$ _____

4. $\dfrac{25}{5} =$ _____ 5. $\dfrac{27}{9} =$ _____ 6. $\dfrac{24}{4} =$ _____

Use digits and three different symbols to show:

7. thirty divided by six $\overline{)}$ $\div$ ___

8. fifty-four divided by nine $\overline{)}$ $\div$ ___

9. Use the numbers 7, 42, and 6 to write two multiplication facts and two division facts.

_____ × _____   _____ × _____

$\overline{)}$ $\overline{)}$

Name _____

Reteaching

Lesson 48

• **Multiplying Two-Digit Numbers, Part 2**

- When multiplying two-digit numbers, sometimes the product of the ones digits is a two-digit number.
- We can use mental math or pencil and paper to carry tens into the tens column.
- To multiply two-digit numbers:
  1. Multiply by the ones digit.
  2. If the product is a two-digit number, write the last digit in the ones column.
  3. Carry the first digit into the tens column.
  4. Multiply by the tens digit.
  5. Add the carried digit to that product.
  6. Write the sum in the anwer line.

Example:

1. 7 × 2 = 14
2. Write 4 in the ones column.
3. Carry the 1.

4. 7 × 3 = 21
5. Add the carried 1 to that product (21 + 1 = 22).
6. Write the 22.

Practice:

Find each product using mental math or pencil and paper to carry. Remember to write the dollar sign in money problems.

1. 17
   × 4

2. 26
   × 8

3. $35
   × 5

4. 51
   × 3

5. $96
   × 6

6. 74
   × 2

7. 65
   × 8

8. $49
   × 9

9. 68
   × 7

Name _____

Reteaching 49
Lesson 49

- **Word Problems about Equal Groups, Part 1**

- Some problems involve equal groups.
- Problems about equal groups follow a multiplication formula. We can find the total by multiplying the number of groups by the number of things in each group.

```
     Number in each group
   × Number of groups
     Total
```

Number of groups × Number in each group = Total

Example: Portia has 6 bottles of water. There are 12 ounces of water in each bottle. How many total ounces of water does Portia have?

Solution:
```
   12 ounces in each bottle      number in each group
 ×  6 bottles of water            number of groups
   72 ounces of water             total
```

Practice:

1. There are 7 days in one week. How many days are in 6 weeks?

   ```
              _____ days in a week
            × _____ weeks
   ```

   _____ days

2. It takes four push pins to hang a poster. There are 9 posters in the classroom. How many push pins will it take to hang all of the posters?

   ```
            × _____
   ```

   _____ push pins

3. A recipe makes 1 batch of biscuits. If the recipe calls for 4 eggs, how many eggs will be needed to make 5 batchs of biscuits?

   ```
            × _____
   ```

   _____ eggs

Name _____

Reteaching 5
Lesson 50

- **Adding and Subtracting Decimal Numbers, Part 2**

  - Decimal place value from **hundreds** to **hundredths**:

    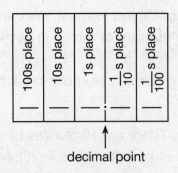

    decimal point

  - To add or subtract decimal numbers:
    1. Line up the decimal points first.
    2. Put zeros in empty spaces.
    3. Add or subtract.

  - Sometimes we need to add or subtract decimal numbers that do not have the same number of decimal places.

    Examples: 1.28 + 3.35 + 2.40   6.37 − 4.9

    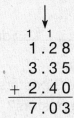

    Line up the decimals   Line up the decimals

    ```
      1 1
      1.28                    5 13
      3.35                    6.3̸7̸
    + 2.40                  − 4.90    ← Use a zero as
      7.03                    1.47       a placeholder.
    ```

Practice:

1. Which digit in 22.6 is in the tenths place? _____

2. Which digit in 115.73 is in the hundredths place? _____

3. Which digit in 10.2 is in the same place as the 9 in 1.91? _____

Find each sum or difference.

4. 4.16 + 2.8 =   5. 4.16 − 2.8 =   6. 13.99 + 4.58 =

```
    4.16              4.16                 13.99
  +  .              −  .                 +   .
  _____            _____               _____
```

54   © Harcourt Achieve Inc. and Stephen Hake. All rights reserved.   Saxon Math Intermediate 4

Name _____

Reteaching Inv. 5
Investigation 5

- **Percents**

  - Think about fractions as percents. A whole is 100%.

  $\frac{1}{4} = 25\%$ $\frac{1}{2} = 50\%$ $\frac{3}{4} = 75\%$ $1 = 100\%$

  - To estimate percents, find out if the amount shaded is more or less than a half. Which choices can you rule out?

  A 30%   **B** 50%   **C** 70%   **D** 90%

We can rule out choice A and B because we know the glass is more than half full but not all the way full. To determine if the answer is **C** or **D**, we must determine if the glass is closer to 50% full or 100% full. From the picture we see the glass is closer to 70% full.

The answer is **D**.

Practice:

1. If 75% of the flowers bloomed this year, what percent of the flowers did not bloom?

2. Compare: 27% $\frac{1}{4}$

3. Compare: 95% $\frac{1}{1}$

4. Which percent best describes the shaded portion of this circle?

   A 10%   **B** 25%   **C** 50%   **D** 80%

Name _____   **Reteaching 5**

Lesson 51

- **Adding Numbers with More Than Three Digits**
- **Checking One-Digit Division**

Adding Numbers with More Than Three Digits
- When writing whole numbers in columns, carefully line-up digits starting with the ones digit in each number.

    Example: 467 + 589 + 1060 + 23

    ```
       467
       589
      1060
    +   23
    ```

Checking One-Digit Division
- We can check a division answer by multiplying the numbers outside the division box.

    Example: $3\overline{)15}^{\,5}$ Check: 5 × 3 = 15

Practice:

Add.

1.   1234
   + 607

2.   47,019
   + 21,598

3.   405,679
   + 319,477

Divide. Check each answer by multiplying.

4. $3\overline{)24}$ ☐
              × 3

5. $7\overline{)49}$ ☐
              × 7

6. $6\overline{)54}$ ☐
              × 6

7. $9\overline{)27}$ ☐
              × 9

Name _____

Reteaching 52

Lesson 52

- **Subtracting Numbers with More Than Three Digits**
- **Word Problems About Equal Groups, Part 2**

Subtracting Numbers with More Than Three Digits

- Always start subtracting in the ones column. Then continue subtracting from right to left.

    Example: 1157 − 1080

    ```
      1 1 5 7        1 1̸ 5̸ 7        1 1̸ 5̸ 7        1 1̸ 5̸ 7
    − 1 0 8 0      − 1 0 8 0      − 1 0 8 0      − 1 0 8 0
    ─────────      ─────────      ─────────      ─────────
            7            7 7          0 7 7        0 0 7 7
    ```
    (with 0 15 shown above the 1 and 5)

Word Problems About Equal Groups

- In word problems, the word "each" usually means an equal-groups problem.
- To find the number in each group, when given the total, we can divide by the number of groups.

$$\text{number of groups} \overline{)\text{total}}^{\text{number in each group}}$$

Practice:

Subtract.

1.   1234
   − 607

2.   47,019
   − 21,598

3.   405,679
   − 319,477

4. There are 48 people. There are 6 equal teams. How many people are in each team?

)_____ _____ people in each team

5. Thirty-six students lined up equally on four risers for a chorus recital. How many students were on each riser?

)_____ _____ students on each riser

Name _____

Reteaching
Lesson 53

- **One-Digit Division with a Remainder**

  - Sometimes when we try to divide a number of things into **equal groups** we have some things left over.

    Example: These 11 triangles cannot be divided into equal groups of four, because there are 3 triangles left over.

    11 triangles   2 groups of four triangles   3 triangles left over

  - We call the amount left over the **remainder**. Use the letter R to identify the remainder in an answer.

    $$\begin{array}{r} 2\ R3 \\ 4\overline{)11} \\ -8 \\ \hline 3 \end{array}$$

  - Any remainder in a division problem must be smaller than the **divisor**.

Practice:

1. Circle groups of triangles below to show 18 ÷ 4. Write the answer shown by your sketch.

   _____ R ____

Divide. Write each answer with a remainder.

2. $2\overline{)15}^{\ \ R}$ 3. $5\overline{)13}^{\ \ R}$

4. $4\overline{)21}^{\ \ R}$ 5. 17 ÷ 2 → $\overline{)17}^{\ \ R}$

6. 27 ÷ 6 → $\overline{)27}^{\ \ R}$ 7. 20 ÷ 3 → $\overline{)20}^{\ \ R}$

Name _____

Reteaching 54

Lesson 54

- **The Calendar**
- **Rounding Numbers to the Nearest Thousand**

The Calendar
- A **common year** has 365 days.
- A **leap year** has 366 days. The extra day is added to February. A leap year happens every 4 years.
- This will help you remember how many days are in each month:

  Thirty days have September, April, June, and November.
  The other months have 31 days, except February,
  which has 28, or 29 if it is leap year.

- A **decade** is ten years. A **century** is one hundred years.
- To find the amount of time between two years, subtract.

  $$\begin{array}{r} 1996 \\ -\ 1983 \\ \hline 13 \end{array}$$ years

Rounding Numbers to the Nearest Thousand
- To round a number to the nearest thousand:
    1. Look at the hundreds place.
    2. Ask: Is the digit in the hundreds place 5 or more? (5, 6, 7, 8, 9)
       Yes → Add 1 to the thousands place.
       No → The thousands place stays the same.
    4. Replace the numbers after the thousands place with zeros.

  Example: 6<u>2</u>59 → 6000

Practice:

Remember to write the units.

1. How often does a leap year occur? _____

2. According to this calendar, what is the date of the third Wednesday of the month?

   ____ ____ / ____ ____ / ____ ____ ____ ____

3. How many years were there from 1913 to 1958? _____

Round to the nearest thousand.

4. 7901 _____   5. 3399 _____

MAY 2014

S	M	T	W	T	F	S
				1	2	3
4	5	6	7	8	9	10
11	12	13	14	15	16	17
18	19	20	21	22	23	24
25	26	27	28	29	30	31

Name _____

Reteaching 55
Lesson 55

- **Prime and Composite Numbers**

  - Multiples are the numbers we say if we count by a number. For example, the multiples of 4 are: 4, 8, 12, 16, 20, 24, ...
  - You can find multiples in a multiplication table.
  - To find the **factors** of a whole number:
    1. Start with the number 1.
    2. End with the number given.
    3. Find all the numbers that divide evenly into the given number:
       Will 2 divide evenly?
       Will 3 divide evenly? (and so on)
    4. Make sure the factors are listed in order.

    Example: List the factors of 30. _1_ , _2_ , _3_ , _5_ , _6_ , _10_ , _15_ , _30_

  - Counting numbers that have exactly two different factors are **prime numbers**.
  - A number with more than two factors is a **composite number**.
  - The number 1 has one factor and is not prime or composite.

Practice:

1. Write all the prime numbers less than 12. _____

2. What is the eighth multiple of 3? _____

3. Is the last digit of the multiples of 4 odd or even? _____

4. List the six factors of 12. _____

5. List the factors of 16.

   _____ , _____ , _____ , _____ , _____

6. Two factors of 20 are 1 and 20. Find four more factors of 20.

   _____ , _____ , _____ , _____

7. List the factors of 11. _____ , _____

8. List the factors of 24. _____ , _____ , _____ , _____ ,

   _____ , _____ , _____

Name _____

Reteaching 56

Lesson 56

- # Using Models and Pictures to Compare Fractions

  - When we draw pictures to compare fractions, the pictures must have the same shape and equal size. These are called **congruent figures**.
  - Another way to compare fractions:
    1. Cross multiply.
    2. Compare the products.

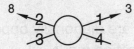

$8 > 3$, so $\frac{2}{3} > \frac{1}{4}$

Practice:

Compare the fractions and shade the rectangles to illustrate each comparison. Use fraction manipulatives for help.

1. $\frac{3}{5} \bigcirc \frac{2}{3}$

2. $\frac{1}{4} \bigcirc \frac{2}{5}$

3. $\frac{3}{5} \bigcirc \frac{1}{3}$

4. $\frac{5}{7} \bigcirc \frac{4}{5}$

5. $\frac{3}{8} \bigcirc \frac{2}{7}$

6. $\frac{4}{9} \bigcirc \frac{5}{8}$

Saxon Math Intermediate 4 © Harcourt Achieve Inc. and Stephen Hake. All rights reserved.

Name _____ **Reteaching 5**
Lesson 57

• **Rate Word Problems**

Example: If a cyclist rides 15 miles per hour, how far will he ride in 6 hours?

1. Name the two things the problem is about:
   miles
   hours

2. Fill in what you know and what you are looking for: $\dfrac{\text{miles}}{\text{hour}} \dfrac{15}{1} = \dfrac{?}{6}$

3. Draw a loop around the numbers that are diagonally opposite. The loop should never include the question mark.

   miles   (15 = ?
   hour   1)   6

4. Multiply the numbers inside the loop and divide by the number outside the loop if it is not 1:   6 × 15 = 90 miles

---

Practice:

1. Maya drove 65 miles in one hour. At that rate, how far can she drive in 7 hours?

   Multiply the loop.

   miles
   hours

   _____ miles

2. Kirby could type 42 words in 3 minutes. At that rate, how many words could he type in 30 minutes?

   Multiply the loop.

   words
   minutes

   _____ words

3. Emma is the fastest runner in her class. She can run 2 miles in 15 minutes. At that rate, how many minutes would it take her to run a 6-mile race?

   Multiply the loop.

   miles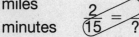
   minutes

   _____ minutes

Name _____

Reteaching 58
Lesson 58

- # Multiplying Three-Digit Numbers

  - We can multiply three-digit numbers the same way we multiplied two-digit numbers: one digit at a time.

    Example 1:

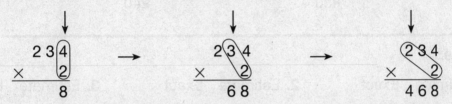

  - Try to use mental math to carry tens.

    Example 2:

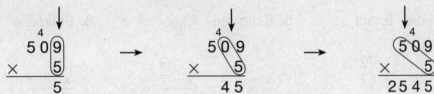

Practice:

Multiply. Remember to write the dollar sign in money problems.

1. 248
    ×   4

2. $618
    ×   5

3. $3.87
    ×   7

4. 501
    ×   6

5. $117
    ×   9

6. $8.34
    ×   3

7. $4.39
    ×   7

8. 723
    ×   8

9. 916
    ×   4

Saxon Math Intermediate 4

Name _____

Reteaching
Lesson 59

- **Estimating Arithmetic Answers**

  - To estimate an answer we often round numbers first.
  - When we estimate, we find an answer that is "close to" the exact number.
  - Estimating can help you see whether your exact answers make sense.

  $$\begin{array}{r}486\\+319\end{array} \rightarrow \begin{array}{r}500\\+300\\\hline 800\end{array} \qquad \begin{array}{r}64\\\times\ 4\end{array} \rightarrow \begin{array}{r}60\\\times\ 4\\\hline 240\end{array} \qquad 53 \div 5 \rightarrow \begin{array}{r}10\\5\overline{)50}\end{array}$$

Practice:

1. Estimate   Exact

$$\begin{array}{r}61\\68\\+\underline{}\end{array} \qquad \begin{array}{r}61\\68\\+\ 47\end{array}$$

2. Estimate   Exact

$$+\underline{} \qquad \begin{array}{r}519\\+\ 354\end{array}$$

3. Estimate   Exact

$$-\underline{} \qquad \begin{array}{r}473\\-\ 250\end{array}$$

4. Estimate   Exact

$$-\underline{} \qquad \begin{array}{r}72\\-\ 67\end{array}$$

5. Estimate   Exact

$$\times\underline{} \qquad \begin{array}{r}39\\\times\ 7\end{array}$$

6. Estimate   Exact

$$\times\underline{} \qquad \begin{array}{r}465\\\times\ 8\end{array}$$

7. Estimate   Exact

$$\overline{)} \qquad 4\overline{)63}$$

8. Estimate   Exact

$$\overline{)} \qquad 6\overline{)55}$$

9. Carlos estimated the product of 6 and 6384 by multiplying 6 by 6000. Was Carlos' estimate more than, equal to, or less than the actual product? Why?

Carlos' estimate was _____ the actual product because

he rounded 6384 down to _____ before multiplying.

Name _____

Reteaching 60

Lesson 60

- **Rate Problems with a Given Total**

  - Rate problems are equal-group problems. To find a missing number in an equal-groups problem (when the total is given), we can divide.

    Example: Marquez can read 4 pages in 1 minute. How long will it take him to read 32 pages?

    $$\text{known number} \overline{)\text{total}}^{\text{missing number}} \qquad 4\text{ pages}\overline{)32\text{ pages}}^{\,8\text{ minutes}}$$

    8 minutes to read 32 pages

Practice:

1. Samantha can sign 15 thank-you cards in 1 minute. How long will it take Samantha to sign cards for the 45 people in her dance troop?

    $\overline{)}$ _____ minutes

2. Farley went skiing with his family. If he travels at 6 feet per second, how long will it take him to travel 48 feet?

    $\overline{)}$ _____ seconds

3. Destiny makes $7 an hour working at the animal shelter. If her paycheck at the end of the day is $42, how many hours did she work?

    $\overline{)}$ _____ hours

Name _____

Reteaching
Investigation 6

- **Displaying Data Using Graphs**

  - A **survey** is an effort to gather specific information about a group, or a population.
  - A **pictograph** uses pictures to display information.
  - A **bar graph** displays numerical information with shaded rectangles or bars.
  - A **line graph** displays numerical information as points connected by line segments. Line graphs are often used to show information that changes over time.
  - **Circle graphs** or pie graphs are often used to display information about parts of a whole.
  - A **legend** is often shown on a graph to describe the meaning of symbols.

Example:

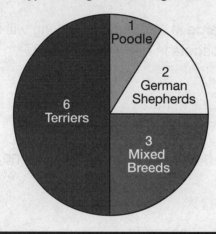

Type of Dogs at the Dog Park

Practice:

Use the circle graph in the example to answer problems 1–3.

1. How many German Shepherds and Poodles are at the Dog Park? _____

2. What is the total number of dogs represented by the circle graph? _____

3. Which type of dog does the largest slice of the graph represent? _____

4. Create a bar graph to represent the same information as the circle graph.

5. Is it easier to read the results from the bar graph or the circle graph? Explain why.

_____

66 © Harcourt Achieve Inc. and Stephen Hake. All rights reserved. *Saxon Math* Intermediate 4

Name _____

Reaching 61

Lesson 61

- **Remaining Fractions**
- **Two-Step Equations**

Remaining Fractions
- If a whole has been divided into parts and we know the size of one part, then we can figure out the size of the other parts.
    - What fraction of the circle is shaded?
    - What fraction of the circle is not shaded?

Two-Step Equations
- To solve a two-step equation:
    1. Find the answer to the right-hand side of the equation.
    2. Find the number for n.

Example:

$2n = 8 + 6$
$2n = 14$ The answer to the right-hand side is 14.
$n = 7$ The number n is 7.

- There is more than one way to show multiplication:

    2×5 The times sign
    $2n$ A number followed by a letter
    $2 \cdot 5$ A raised dot

Practice:

Use fraction manipulatives for help.

1. What fraction of this rectangle is not shaded? _____

2. Two-fifths of the show was over.
   What fraction of the show was left? _____

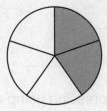

Find each missing number.

3. $5n = 7 + 8$ $n =$ _____   4. $3n = 12 + 6$ $n =$ _____

5. $4n = 4 + 4$ $n =$ _____   6. $2w = 4 \cdot 4$ $w =$ _____

Name _____

Reteaching
Lesson 62

- **Multiplying Three or More Factors**
- **Exponents**

Multiplying Three or More Factors
- To find the product of three numbers:
    1. Multiply any two of the numbers.
    2. Multiply that answer by the third number.
- It does not matter which numbers are multiplied first.

$1 \times 2 \times 3 =$ $1 \times 2 \times 3 =$
$2 \times 3 = 6 \times 1 = 6$ $1 \times 2 = 2; 2 \times 3 = 6$

Exponents
- An **exponent** is a number that shows how many times another number (the **base**) is to be used as a factor. An exponent is written above and to the right of the base.

    base → 4^2 ← exponent
    4^2 means 4×4.
    4^2 equals 16.

- If the exponent is 2, we say "squared" for the exponent.
  So 4^2 is read as "four squared."
- If the exponent is 3, we say "cubed" for the exponent.
  So 4^3 is read as "four cubed."

Practice:

Simplify.

1. $2 \times 4 \times 6 =$ _____
2. $4 \times 5 \times 8 =$ _____
3. $6^2 =$ _____
4. $5^3 =$ _____
5. $8^2 - 6^2 =$ _____
6. $3^3 + 2^3 =$ _____

Rewrite the expression using an exponent.

7. $3 \times 3 \times 3 \times 3 \times 3 =$ _____
8. $5 \times 5 \times 5 \times 5 \times 5 \times 5 \times 5 =$ _____

Name _____

Reteaching 63
Lesson 63

- **Polygons**

  - **Polygons** are closed, flat shapes made from line segments.
  - Each corner of a polygon is called a **vertex**. The plural of vertex is vertices.
  - **Regular** polygons have **sides** of equal **length** and **angles** of equal measure.

Three-sided polygons are **triangles**.	(**Regular** triangle)
Four-sided polygons are **quadrilaterals**.	(A square is a **regular** quadrilateral.)
Five-sided polygons are **pentagons**.	(**Regular** pentagon)
Six-sided polygons are **hexagons**.	(**Regular** hexagon)
Eight-sided polygons are **octagons**.	(**Regular** octagon)
Ten-sided polygons are **decagons**.	(**Regular** decagon)

Practice:

Draw an example of each polygon.

1. regular triangle
2. hexagon
3. decagon

Name _____

Reteaching 64
Lesson 64

• Division with Two-Digit Answers, Part 1

- In a division problem, the number being divided is called the **dividend**. The dividend is divided by a **divisor**. The answer is called the **quotient**.

$$\text{divisor}\overline{)\text{dividend}}^{\text{quotient}} \qquad \text{dividend} \div \text{divisor} = \text{quotient}$$

- We can break a division problem with a two-digit dividend into easier steps: divide, multiply, subtract, and "bring down".
- First, test for divisibility.

Tests for Divisibility	
A number is able to be divided by:	
2 if the last digit is even.	⎫
5 if the last digit is 0 or 5.	⎬ Already know
10 if the last digit is 0.	⎭
3 if the sum of the digits can be divided by 3.	} New

Example: 96 can be divided by 3 with no remainder because
9 + 6 = 15   and   15 is a multiple of 3.

Practice:

Practice the division steps to solve these problems.

1. $2\overline{)54}$
2. $5\overline{)65}$

3. $3\overline{)75}$
4. $7\overline{)91}$

5. $6\overline{)84}$
6. $8\overline{)96}$

7. $3\overline{)81}$
8. $4\overline{)76}$

9. Which of these numbers can be divided by 3 with no remainder? _____

   A 46   **B** 72   **C** 53   **D** 61

   How do you know? *The sum of* _____ *and* _____ *is 9, which is a multiple of 3.*

Name _____

Reteaching 65

Lesson 65

- **Division with Two-Digit Answers, Part 2**

    - The numbers in a division problem are named the **divisor**, **dividend**, and **quotient**.

$$\text{divisor} \rightarrow \quad 4\overline{)64} \quad \leftarrow \text{dividend} \\ \phantom{\text{divisor} \rightarrow \quad } 16 \leftarrow \text{quotient}$$

    - First, test for divisibility.

Tests for Divisibility
A number can be divided by:
2   if the last digit is even.
5   if the last digit is 0 or 5.
10 if the last digit is 0.
3   if the sum of the digits can be divided by 3.
9   if the sum of the digits can be divided by 9.

Already know (for 2, 5, 10, 3) } New (for 9)

    - To divide, follow the division steps (divide, multiply, subtract, "bring down") for each place in the dividend. This way of dividing is called **long division**.

    Start with the first digit.

    If the digit cannot be divided, use zero as a placeholder in the quotient.

    Place a digit in the quotient above each digit in the dividend.

Practice:

In the division fact 28 ÷ 7 = 4,

1. What number is the divisor? _____

2. What number is the quotient? _____

3. What number is the dividend? _____

Divide.

4. $4\overline{)156}$

5. $6\overline{)156}$

6. $4\overline{)204}$

7. $3\overline{)204}$

Name _____

Reteaching 6
Lesson 66

• Similar and Congruent Figures

- If two figures are **congruent**, their corresponding parts (angles and sides) match exactly.

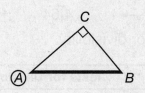

    Example: Triangle ABC and triangle XYZ are congruent.

        ∠A   corresponds to   ∠X

        $\overline{AB}$ corresponds to $\overline{XY}$

- If two figures are **similar**, they have the same shape but not necessarily the same size.
- Similar figures have equal, matching angles.

    Triangles A, B, and C are similar.
    Triangles A and C are congruent.
    Triangle D is not similar or congruent to triangle A, B, or C.

Practice:

Refer to the triangles at right for problems 1 and 2.

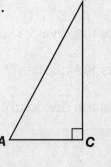
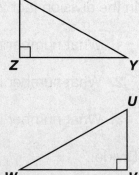

1. Which of the triangles are similar?

_____

2. Which of the triangles are congruent?

_____

Refer to the triangles at right for problems 3 and 4.

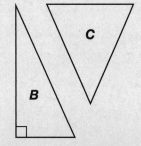
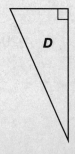

3. Which of the triangles are similar?

_____

4. Which of the triangles are congruent?

_____

Reteaching 67

Lesson 67

- **Multiplying by Multiples of 10**

  - To multiply a whole number by 10, just attach a zero after the number.
    $$23 \times 10 = 230 \qquad 48 \times 10 = 480$$
  - One way to multiply a whole number or a decimal number by a multiple of 10 is to use **offset multiplication**.
    1. Write the multiple of 10 as the bottom number.
    2. Let the zero "hang out" ("offset" to the right).
    3. Copy the zero into the answer.
    4. Multiply.

    ```
      42                                    42        42
    ×  20  ← zero "hangs out" to the right × 20      × 20
                                              0       840
    ```

  - When multiplying a money amount by a multiple of 10, put two decimal places in the answer.

    ```
      $1.34
    ×    20
     $26.80
    ```

Practice:

Multiply. Remember to write the dollar sign in money problems.

1. 65 × 10 = _____

2. 10 × 41 = _____

3. 10 × 78¢ = _____

4.   57
   × 20

5. $2.93
   ×   30

6.   74
   × 40

7. Write 16 × 40 as a product of 10 and two other factors. Then multiply.

   _____ × _____ × 10 = _____

Saxon Math Intermediate 4

Name _____

Reteaching 68
Lesson 68

- **Division with Two-Digit Answers and a Remainder**

  - In **long division**, remember four steps:
    Step 1: Divide.
    Step 2: Multiply.
    Step 3: Subtract.
    Step 4: Bring down.

Division Chart

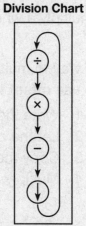

Practice:

Divide. Use **long division**.

1. 3)‾104   ___ R ___

2. 7)‾237   ___ R ___

3. 5)‾172   ___ R ___

4. 6)‾261   ___ R ___

5. 3)‾119   ___ R ___

6. 8)‾396   ___ R ___

7. 9)‾640   ___ R ___

8. 6)‾456   ___ R ___

9. 3)‾100   ___ R ___

10. Deshawn divided 138 by 4 and got 34 R2 for his answer. Describe how to check Deshawn's calculation.

    To check his calculation, I would _____ 34 by _____. Then I would

    _____ 2 to the product.

    The answer should be _____.

74   © Harcourt Achieve Inc. and Stephen Hake. All rights reserved.   *Saxon Math* Intermediate 4

Name _____

Reteaching 69
Lesson 69

- **Millimeters**

  - It takes 10 millimeters to equal 1 centimeter.
  - A millimeter scale and a centimeter scale are shown below.

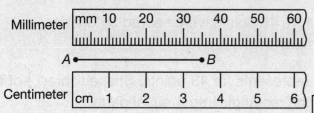

  - The length of segment AB is 35 millimeters. It is also 3.5 centimeters.
  - To convert between millimeters and centimeters, multiply or divide by 10.

Convert:
1 cm = 10 mm
3 cm = _____ mm
5 cm = _____ mm
_____ cm = 20 mm
_____ cm = 40 mm

  - **Metric** measures are always written as **decimal** numbers instead of fractions.

Practice:

1. The width of a fingernail is about 1 centimeter. How many centimeters is 3 meters? _____

2. A dime is about 1 millimeter thick. How many dimes would it take to make a stack 5 centimeters high? _____

3. Each side of a triangle is 4 centimeters long. What is the perimeter of the triangle? _____

4. The diameter of a quarter is about 23 mm. How many centimeters is that? _____

5. A rectangle has a length of 6 cm and a width that is half that. What is the perimeter in millimeters? _____

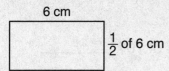

Saxon Math Intermediate 4   © Harcourt Achieve Inc. and Stephen Hake. All rights reserved.   75

Name _____

Reteaching
Lesson 70

- **Word Problems About a Fraction of a Group**

- To find the fractional part of a group we divide by the number of equal parts. To divide, we use the denominator.

    Example: One fourth of the team's 48 points were scored by Shane. Shane scored how many points?

    Solution: The whole rectangle represents for 48 points. Shane scored $\frac{1}{4}$ of the points, so we divide the rectangle into 4 equal parts.

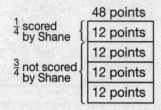

$$\frac{12}{4\overline{)48}}$$

Shane scored 12 points

Shortcuts:

$\frac{1}{2}$ of a number   Divide by 2.

$\frac{1}{3}$ of a number   Divide by 3.

$\frac{1}{4}$ of a number   Divide by 4.

Practice:

1. What is $\frac{1}{3}$ of 27? _____

2. What is $\frac{1}{2}$ of 14? _____

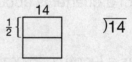

3. What is $\frac{1}{4}$ of 48? _____

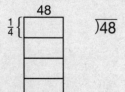

4. What is $\frac{1}{5}$ of 70? _____

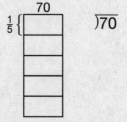

Name _____

Reteaching Inv. 7
Investigation 7

- **Collecting Data with Surveys**

  - The students who answer the questions are participating in a **survey.**
  - People that are a part of the survey are giving information about a group or a population. This part is called a **sample** of the population.
  - When conducting surveys, we can use **tally marks** to record responses. Each tally mark represents one time the event or outcome occurred.

Practice:

Answer the questions below using the tally sheet shown.

Student Shoe Color

Color	Tally
White	|||| ||||
Blue	||||	
Black	|||| |||| ||||	
Other	|||| ||||

1. How many students are wearing white shoes? _____

2. How many more black shoes are there than blue shoes? _____

3. Which color of shoe are the most of the students wearing? _____

4. The number of black shoes is equal to the number of which two colors of shoes combined? _____

5. What does the category "Other" represent? _____

   _____

6. Which color of shoes are you wearing? Add a tally mark to the tally sheet above to represent your shoe color. _____

Saxon Math Intermediate 4   © Harcourt Achieve Inc. and Stephen Hake. All rights reserved.   **77**

Reteaching 7

Lesson 71

- **Division Answers Ending with Zero**

  - Sometimes the last whole digit in a division answer is zero.
  - Continue following the division steps until there are no more digits in the dividend to divide.
  - If the digit in the dividend cannot be divided, place a zero in the quotient above the digit.

Practice:

1. 3)̄122   ⁰ R

2. 5)̄204   ⁰ R

3. 6)̄185   ⁰ R

4. 3)̄242   ⁰ R

5. 4)̄360   R

6. 4)̄83   R

7. 5)̄152   R

8. 6)̄304   R

Name _____   **Reteaching 72**
                                                   Lesson 72

- **Finding Information to Solve Problems**

  - Some word problems might contain more information than you need to solve the problem.
  - Read all word problems carefully.

Practice:

Read the word problems. Then answer the questions that follow. Remember to write the units.

Bernard went hiking on Sunday. He hiked for five hours from the trail entrance to Packsaddle Falls. It took Bernard four hours to hike back to the trail entrance. From the trail entrance to Packsaddle Falls is 10 miles.

1. How many hours did Bernard walk in all? _____

2. About how long did it take Bernard to walk one mile on the way to Packsaddle Falls? _____

3. How many miles did Bernard walk altogether? _____

Charisma went to the mall to buy birthday gifts for her twin cousins. She bought a video game for $32. For her other cousin, she bought a scooter. Charisma spent $89 altogether.

4. How much did Charisma spend on the scooter? _____

5. How much more did it cost than the video game? _____

6. If Charisma had bought two video games for $38 each and a scooter, how much would she have spent? _____

Taryn's choir group is planning a trip for the state competition. The bus ride will take 8 hours each way. When they arrive they will wait for 2 hours before performing. Their performance lasts 1 hour. They will eat dinner and return to the bus 2 hours after their performance.

7. How many hours does Taryn's choir spend on the bus altogether? _____

8. How many hours do they spend at the competition? _____

9. How many total hours is the trip? _____

Saxon Math Intermediate 4

Name _____

Reteaching 73
Lesson 73

• Geometric Transformations

• Slides, turns, and flips are three ways of moving figures. In geometry, we call these movements **transformations**.

Transformations	
Movement	Name
Slide	Translation
Turn	Rotation
Flip	Reflection

Example: Position triangle ABC on triangle XYZ using the three transformations.

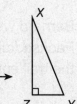

Solution:

1. Rotation: **Turn** triangle ABC about point C.

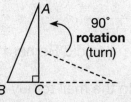 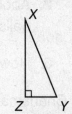

2. Translation: **Slide** triangle ABC to the right.

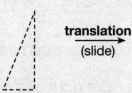 

3. Reflection: **Flip** triangle ABC over line AC.

Practice:

1. What is the math term for a turn? _____

2. What is the math term for a slide? _____

3. What is the math term for a flip? _____

Name _____

Reteaching 74

Lesson 74

• **Fraction of a Set**

- A set is a group of similar things.
- We can use fractions to describe a number of things in a set (group).

    Example:

       $\frac{4}{7}$ Four circles are shaded.
    There are seven circles in all.
    Four out of seven are shaded.

Practice:

1. What fraction of the set is shaded?

2. What fraction of the set is not shaded?

 _____

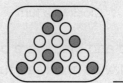 _____

3. What fraction of the set is shaded?

4. What fraction of the set is not shaded?

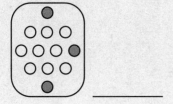 _____

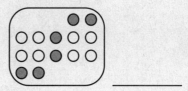 _____

5. There are 29 students in class. There are 16 girls and 13 boys. What fraction of the class is boys? _____

6. What fraction of the letters in ONOMOTOPOEIA are Os? _____

7. When preparing for the photography competition, Richard took 61 pictures. If 32 were color pictures, what fraction of the pictures were black and white?
_____

8. Meredith was making bracelets for her friends. She made 17 bracelets, 8 of which were for her friends in art. What fraction of the bracelets are for friends not in Meredith's art class? _____

Name _____

Reteaching 75
Lesson 75

- **Measuring Turns**

  - Turns can be measured in **degrees**.

    A **full turn** is 360°.
    A **half turn** (a turn to face the opposite direction) is 180°.
    A **quarter turn** is 90°.

  - **Clockwise** turns go in the same direction as the hands of a clock.
  - **Counterclockwise** turns go in the opposite direction as the hands of a clock.

Clockwise turn   Counterclockwise turn

Example: Todd and Matt were both facing north. Todd turned 90° clockwise and Matt turned 90° counterclockwise. After turning, in which directions were the boys facing?

Solution: Todd was facing east and Matt was facing west. Below are the turns Todd and Matt made.

```
      N  90°              90°  N
W  →  E              W  ←      E
      S                    S
   Todd                  Matt
 Clockwise           Counterclockwise
```

Practice:

1. Sara walked out the door. Then she turned and walked back into the house to get a jacket. About how many degrees did she turn?

2. Rolf walked north on First Avenue, then turned right on Lemon Street. Did Rolf turn clockwise or counterclockwise? About how many degrees did he turn?

82   © Harcourt Achieve Inc. and Stephen Hake. All rights reserved.   Saxon Math *Intermediate 4*

Name _____

Reteaching 76

Lesson 76

- **Division with Three-Digit Answers**

- With long division, remember there are four steps: divide, multiply, subtract, and "bring down."

$$\begin{array}{r} 223 \text{ R3} \\ 4\overline{)89^15} \end{array}$$

- Continue following the division steps until there are no more digits in the dividend to divide. Write any amount "left over" as the remainder.

Practice:

Divide.

1. $4\overline{)967}$ ___ R___

2. $5\overline{)\$8.65}$ $$ ___.___ R___

3. $6\overline{)1814}$ ___0___ R___

4. $8\overline{)\$70.00}$ $0___.___ R___

5. $4\overline{)3402}$ ___0___ R___

6. $3\overline{)200}$ ___ R___

7. $2\overline{)1111}$ ___ R___

8. $6\overline{)7017}$ ___ R___

Saxon Math Intermediate 4 © Harcourt Achieve Inc. and Stephen Hake. All rights reserved.

Name _____

Reteaching 77
Lesson 77

- **Mass and Weight**

U.S. Customary Units of Weight
- The units of weight in the U.S. Customary System are **ounces** (oz), **pounds** (lb), and **tons** (tn).

$$16 \text{ oz} = 1 \text{ lb}$$
$$2000 \text{ lb} = 1 \text{ ton}$$

A box of cereal weighs about 24 ounces.
Some students weigh 98 pounds.
A car might weigh a ton or more.

Metric Units of Mass
- **Grams** (g) and **kilograms** (kg) are metric units of mass.

$$1000 \text{ g} = 1 \text{ kg}$$

Your textbook has a mass of about 1 kilogram.
One dollar bill has a mass of about 1 gram.

Practice:

Remember to write the units.

1. An adult elephant can weigh up to 12,000 pounds. How many tons is that?

   _____

2. Beka weighed 7 pounds when she was born. How many ounces is 7 pounds?

   1 lb = 16 oz   7 lb = _____ oz

3. What would be the most reasonable measurement for a handfuls of raisins?

   A 30 oz   **B** 30 g   **C** 30 kg   **D** 30 lbs

4. One penny has a mass of about 2 grams. What would be the approximate mass of a stack of 50 pennies? _____

Name _____

Reteaching 78

Lesson 78

- **Classifying Triangles**

Classifying Triangles						
by <u>Sides</u>			by <u>Angles</u>		
Type	Characteristic	Example	Type	Characteristic	Example
Equilateral triangle	Three sides of equal length	△	Acute triangle	All acute angles	△
Isosceles triangle	At least two equal sides	▷	Right triangle	One right angle	◿
Scalene triangle	Three sides of unequal length	△	Obtuse triangle	One obtuse angle	◺

Practice:

1. Can a triangle have only two angles? Why or why not? _____

   _____

2. What is the name for a triangle that has three unequal sides?

   _____

3. What is the name for a triangle that has three equal sides? _____

4. What is the name for a triangle that has a 110° angle? _____

5. What is the name for a triangle that has a 90° angle? _____

6. If one side of an equilateral triangle is 8 inches, what is the perimeter of the triangle? Remember to write the units.

   _____

7. If one side of a scalene triangle is 6 inches, can we find its perimeter? Why or why not?

   _____

8. What do the marks on the sides of the triangle mean? _____

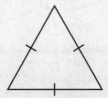

Saxon Math Intermediate 4 © Harcourt Achieve Inc. and Stephen Hake. All rights reserved.   85

Name _____

Reteaching 79

Lesson 79

- **Symmetry**

  - A **line of symmetry** divides a figure in half so that the halves are mirror images of each other.

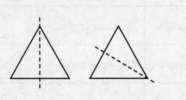 

  - A polygon or other figure can have more than one line of symmetry.

Practice:

Draw lines of symmetry, if any. Write "none" if there are no lines of symmetry.

1.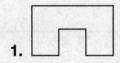
2.
3.

4.
5.
6.

7.
8.
9.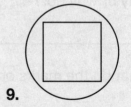

Name _____

Reteaching 80

Lesson 80

- **Division with Zeros in Three-Digit Answers**

- Use the division steps (divide, multiply, subtract, "bring down") as you did for division with zeros in the two-digit answers, but continue until every digit in the dividend is used.
- Place a digit above each digit.
- Use zero as a placeholder.
- Any amount "left over" becomes the remainder.
- Use mental math to divide when possible.

$$8\overline{)3200} \quad \text{Think:} \quad 8\overline{)32}^{4}, \quad \text{so} \quad 8\overline{)3200}^{400}$$

Practice

Divide.

1. $3\overline{)542}$ ___ R ___

2. $3\overline{)6019}$ ___ R ___

3. $4\overline{)2671}$ __0_ R ___

4. $4\overline{)303}$ _0_ R ___

5. $5\overline{)4122}$ _0_ R ___

6. $6\overline{)1991}$ _0_ R ___

Use mental math to divide.

7. $3\overline{)1800}$

8. $3\overline{)2700}$

9. $3\overline{)45{,}000}$

Saxon Math *Intermediate 4*

Name _____

Reteaching Inv.
Investigation 8

• **Analyzing and Graphing Relationships**

  • We can use tables and graphs to show the relationship between two quantities.
  • We use dots to represent the relationship between two quantities on a grid.
  • We use coordinates to show the relationship between two numbers on a grid.

Practice:

1. The table below shows the high temperature for each day of one week. On a separate sheet of paper, create a line graph that shows the information in the table. Place the temperature scale (70° to 85° degrees) on the vertical scale. Place the days of the week (Sunday through Saturday) on the horizontal scale.

Daily High Temperatures

Day	Temperature (°F)
Sunday	70°
Monday	73°
Tuesday	76°
Wednesday	79°
Thursday	82°
Friday	85°
Saturday	88°

2. Which day of the week had the highest temperature? _____

3. Which day of the week had the lowest temperature? _____

4. What does the highest dot on your graph represent? _____

5. What does the lowest dot on your graph represent? _____

6. How much did the temperature change each day? _____

7. On the eighth day, if the temperature rose by the same amount again, what would be the temperature (not shown in table)? _____

Reteaching 81

Lesson 81

• **Angle Measures**

Angles	
Type	Measure
Right angle	90°
Obtuse angle	more than 90°, less than 180°
Acute angle	less than 90°
Straight angle	180°
Full circle	360°

right angle, 90°

straight angle, 180°

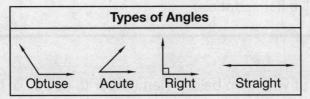

Practice:

Estimate the measure of each angle. Use a protractor to check your estimates. Remember to write the degree symbol.

1.    _____

2. _____

3. _____

4. _____

5.    _____

6. _____

Name _____

Reteaching 82
Lesson 82

- **Tessellations**

  - A **tessellation**, also called a tiling, is the repeated use of shapes to fill a flat surface without gaps or overlaps.

    Examples:

    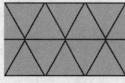  

    triangle   hexagon   quadrilateral

  - Not all polygons tessellate (fill a flat surface), but every triangle and every quadrilateral can fill a flat surface.

Practice:

Create a tessellation of your own below. Remember to use all of the same shape.

Name _____

Reteaching 83
Lesson 83

- **Sales Tax**

- Find the total with **sales tax**. Then subtract to find the change back.

    Example: Ty bought a pair of sneakers priced at $43.99. The sales tax was $2.76. Ty paid the clerk $50.00. How much change should he get back?

    Find the total with tax.

    ```
      $43.99    price of the sneakers
    + $ 2.27    sales tax
      $46.75    total cost
    ```

    Subtract to find the change back.

    ```
        4 9 9 10
      $ 5̸ 0̸.0̸ 0    amount paid
    - $ 4 6.7 5    total cost
      $    3.2 5   change
    ```

Practice:

Remember to write the dollar sign in money problems.

1. Cameron bought lunch at the mall for $6.17. The total sales tax was 89¢. Altogether, how much did Cameron pay for his lunch?

    ```
      $6.17
    + $0.___
      $ .
    ```
    Cameron spent _____ on his lunch.

2. Cicely bought two pairs of jeans for the beginning of school. Each pair was $29.89. The total sales tax was $4.03. How much money did Cicely use to buy her jeans?

    ```
      $29.89         $    .
    + $29.89       + $ 4.03     Cicely spent _____ on jeans.
                     $    .
    ```

3. If Cicely paid for her jeans with four $20 bills, how much change would she receive?

    ```
                        $    .
    $20 × 4 = _____ - $    .
                        $    .
    ```

Saxon Math Intermediate 4

Name _____

Reteaching 84
Lesson 84

- **Decimal Numbers to Thousandths**

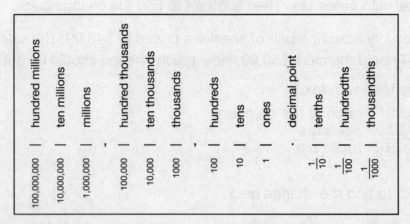

Example: 13.783 is read "thirteen *and* seven hundred eighty-three **thousandths**".

Practice:

Write each fraction or mixed number as a decimal number.

1. $\dfrac{617}{1{,}000}$ ____ . ____ ____ ____

2. $5\dfrac{346}{1{,}000}$ ____ . ____ ____ ____

3. $\dfrac{72}{100}$ ____ . ____ ____

4. $2\dfrac{84}{1{,}000}$ ____ . ____ ____ ____

Write each decimal number as a fraction or mixed number. Then use words to name the numbers.

5. 0.349

   Fraction: _____

   Words: _____
   _____
   _____

6. 5.128

   Mixed Number: _____

   Words: _____
   _____
   _____

7. 0.004

   Fraction: _____

   Words: _____
   _____
   _____

8. 4.405

   Mixed Number: _____

   Words: _____
   _____
   _____

Name _____

Reteaching 85
Lesson 85

- **Multiplying by 10, by 100, and by 1000**

    - If multiplying by 10, add one zero after the digits.
    - If multiplying by 100, add two zeros after the digits.
    - If multiplying by 1000, add three zeros after the digits.

    Example:

    45 × 10 =
    45**0**

- If multiplying money amounts, move the decimal point to the right and use zeros as placeholders. Remember, there must be exactly two decimal places in money amounts.

    Examples:
    38 × 100 = 38**00**   57 × 1000 = 57,**000**
    $5.83 × 100 = $583.00   67¢ × 1000 = $670.00

Practice:

Multiply mentally. Remember to write the dollar sign in money problems.

1. 425 × 10 = _____

2. $5.30 × 10 = _____

3. 67¢ × 100 = _____

4. $3.75 × 100 = _____

5. 6 × 1000 = _____

6. $3.22 × 1000 = _____

7. 54 × 10 = _____

8. 418 × 100 = _____

Name _____

Reteaching
Lesson 86

- **Multiplying Multiples of 10 and 100**

  - To multiply round numbers mentally:
    1. Multiply the non-zero digits of the factors.
    2. Count the zeros in the ones places of the factors.
    3. Attach that number of zeros to the product of the digits.

  Example: 40 × 70

  1. Think, "four times seven equals 28".
  2. There is one zero in 40 plus one zero in 70, or two zeros total.
  3. Attach two zeros to 28.

     40 × 70 = 2800

Practice:

Multiply mentally. Remember to write the dollar sign in money problems.

1. 70 × 80 = _____

2. $5.00 × 4 = _____

3. 60 × 200 = _____

4. $3.00 × 900 = _____

5. 6 × 5000 = _____

6. $4.00 × 3000 = _____

7. 50 × 20 = _____

8. 400 × 500 = _____

Name _____

Reteaching 87
Lesson 87

- **Multiplying Two Two-Digit Numbers, Part 1**

  - Use a three-step process to multiply two two-digit numbers.

    Example: 35
                 × 14

    1. Multiply the top number by the ones digit in the bottom number (ignore the tens digit).

       ²
       35
       ×14
       140

    2. On the next line use a zero as a placeholder in the ones place. Then, multiply the top number by the tens digit in the bottom number.

       35
       ×14
       140
       350

    3. Add the two lines.

       35
       ×14
       140
       +350
       490

Practice:

Multiply.

1.   36
    ×24
    ___
    +   0

2.   24
    ×36
    ___
    +   0

3.   65
    ×17
    ___
    +   0

4.   41
    ×28
    ___
    +   0

Saxon Math Intermediate 4   © Harcourt Achieve Inc. and Stephen Hake. All rights reserved.

Name _____

Reteaching 88

Lesson 88

- **Remainders in Word Problems About Equal Groups**

  - Read word problems carefully to find the information you need to solve the problem.

    Example: The line for juice had 75 people. Juice boxes came in packages of 7.

    a) How many full packages will be needed?
    b) How many more juice boxes are needed?

    $$7\overline{)75} \quad 10 \text{ R } 5$$

    10 full packages would have enough juice boxes for 70 people.
    5 people would still need juice.
    So 11 packages are needed.

Practice:

Mrs. Obije had 49 people enroll for an art class. Each art table can seat a group of 4 students.

1. How many groups of four students will be in the art class? _____ $)\overline{49}$ R

2. How many tables will not be full? _____

3. How many people will be at that table? _____

4. How many total tables will be needed? _____

Samir was having a barbecue at his house. He bought 80 bottles of water. Samir estimated that each of the 26 people he invited would drink about 3 bottles.

5. How many people are able to have 3 bottles of water? _____ $)\overline{80}$ R

6. How many bottles will be left over? _____

7. If 10 of Samir's friends drank 4 bottles of water each, would everyone else be able to have 3? _____

Name _____

Reteaching 89
Lesson 89

- **Mixed Numbers and Improper Fractions**
  - **Improper fractions** are numbers greater than or equal to one.
  - Improper fractions can be converted to mixed numbers:

  Improper ("top heavy") fraction   Whole number and a fraction

  $$\frac{11}{3} = 3\frac{2}{3}$$

- Divide the circles into the number of parts shown in the **denominator**.

Practice:

Use a pencil to shade in the parts of each circle to represent the expression given.

1. $1\frac{4}{5} = \frac{9}{5}$

  =

2. $2\frac{3}{8} = \frac{19}{8}$

   =

3. $1\frac{1}{4} = \frac{5}{4}$

  =

4. $2\frac{1}{3} = \frac{7}{3}$

   =

Name _____

Reteaching

Lesson 90

- **Multiplying Two Two-Digit Numbers, Part 2**

  - When multiplying two two-digit numbers:
    1. Indent the second line and use a zero as a placeholder.
    2. Use mental math to carry (unless the "carried number" is more than 5).
    3. If the "carried number" is more than 5, write it down.

Practice:

1. $\begin{array}{r} 37 \\ \times\ 29 \\ \hline 0 \\ + \\ \hline \end{array}$

2. $\begin{array}{r} 53 \\ \times\ 41 \\ \hline 0 \\ + \\ \hline \end{array}$

3. $\begin{array}{r} 85 \\ \times\ 67 \\ \hline 0 \\ + \\ \hline \end{array}$

4. $\begin{array}{r} 68 \\ \times\ 43 \\ \hline 0 \\ + \\ \hline \end{array}$

5. $\begin{array}{r} 55 \\ \times\ 36 \\ \hline 0 \\ + \\ \hline \end{array}$

6. $\begin{array}{r} 24 \\ \times\ 18 \\ \hline 0 \\ + \\ \hline \end{array}$

Name _____

Reteaching Inv. 9
Investigation 9

- **Investigating Fractions with Manipulatives**

  - A **fraction** is a part of a whole or part of a group.
  - The **numerator** is the top number in a fraction. The **denominator** is the bottom number.
  - The denominator is the total number of equal parts. The numerator is the number of parts described.
  - We read fractions from top to bottom.

    $\frac{1}{2}$ "one half"

    $\frac{1}{4}$ "one fourth" (or "one quarter")

    $\frac{1}{10}$ "one tenth"

  - Fractions, decimals and percents are all related. Each of the amounts above can be represented in multiple ways.

    $\frac{1}{2} = .50 = 50\%$ $\frac{1}{4} = 0.25 = 25\%$ $\frac{1}{10} = 0.10 = 10\%$

Practice:

1. Name the fraction $\frac{1}{5}$. _____

2. How many half-circles make one full circle? Draw circles to show your answer.
   _____

3. What percentage represents a full circle? What percentage represents a half-circle? _____

4. Arrange $\frac{1}{4}$, $\frac{1}{8}$, and $\frac{1}{2}$ from least to greatest. _____

5. What percentage of a circle is $\frac{1}{4}$ a circle? _____

6. What decimal is equivalent to $\frac{1}{4}$? _____

7. Answer this problem using your fraction manipulatives: $\frac{3}{4} - \frac{1}{4} = \frac{\square}{\square}$

Saxon Math Intermediate 4

Name _____

Reteaching 9

Lesson 91

- **Decimal Place Value**

  - A tenth is **greater than** a hundredth.

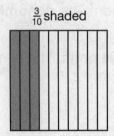
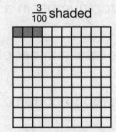

$\frac{3}{10}$ shaded $\frac{3}{100}$ shaded

  - Thinking about money can help us understand decimal place value.

  hundreds   tens   ones   .decimal point   tenths   hundredths

  - To identify decimal place value, pay attention to the decimal point, not the last digit of the number.

  26.73   28.6
  26.730   2.86

  - If there is no digit to subtract from, fill the empty place with a zero.

  Example: 5.1 − 3.38

  ```
      4 10 10
    5. 1̸ 0̸
  − 3. 3  8
    1. 7  2
  ```

Practice:

1. Which digit in 5.186 is in the hundredths place? _____

2. Which digit in 5.186 is in the same place as the 4 in 18.401? _____

3. Name the place value of 6 in the number 3.46. _____

4. Which two numbers below are equal? _____ and _____
   78.95   78.950   7.895

5. Which two numbers below are equal? _____ and _____
   0.020   0.02   0.201

6. Which digit is in the thousandths place in 21.345? _____

Reteaching 92

Lesson 92

- **Classifying Quadrilaterals**

  - A **quadrilateral** is a **polygon** with 4 sides.
  - A **parallelogram** is a quadrilateral with 2 pairs of parallel sides.
  - A **rectangle** is a parallelogram with 4 right angles.
  - A **square** is a rectangle with 4 equal sides.

Compare quadrilaterals by studying the examples in this chart.

Classifying Quadrilaterals		
A quadrilateral is any four-sided polygon.		
Name	Characteristics	Shape
Trapezium	No sides parallel	
Trapezoid	One pair of parallel sides	
Parallelogram	Two pairs of parallel sides	
Rhombus	Parallelogram with equal sides	
Rectangle	Parallelogram with right angles	
Square	Rectangle with equal sides	

Practice:

Describe each quadrilateral as a trapezoid, trapezium, parallelogram, rhombus, rectangle, or square. More than one description might apply to each figure.

1. _____

_____

_____

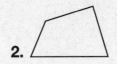

2. _____

_____

_____

3. _____

_____

_____

Name _____

Reteaching
Lesson 93

- **Estimating Multiplication and Division Answers**

  - To estimate multiplication or division answers, round first.

    Examples: 45 × 31   157 ÷ 8

    50 × 30 = 1500   160 ÷ 8 = 20

Practice:

Estimate each product or quotient. Then find the exact answer.

1. 48 × 22

   Estimate: _____ × _____ = _____

   Exact answer:
   ```
        4 8
      × 2 2
      ─────
   +      0
   ```

2. 67 × 54

   Estimate: _____ × _____ = _____

   Exact answer:
   ```
        6 7
      × 5 4
      ─────
   +      0
   ```

3. 29 × 16

   Estimate: _____ × _____ = _____

   Exact answer:
   ```
        2 9
      × 1 6
      ─────
   +      0
   ```

4. 59 × 33

   Estimate: _____ × _____ = _____

   Exact answer:
   ```
        5 9
      × 3 3
      ─────
   +      0
   ```

Name _____

Reteaching 94

Lesson 94

- **Two-Step Word Problems**

  - To solve some word problems we have to perform two operations.
  - Writing down the given information or drawing a picture is often helpful in solving two-step word problems.

Practice:

1. Christa bought 8 bagels with a $10 bill. She got back $6. What was the cost of each bagel? Remember to write the units.

   Write down what you know:

   Christa bought _____. She used _____ to pay for the bagels.

   She got back _____ in change.

   Subtract to find the cost of 8 bagels.

      $10.00
   − $ 6.00

   Divide to find the cost of each bagel.

   _____) _____   *Hint:* Show dollars and cents.

   _____ per bagel

2. The perimeter of this rectangle is 12 inches. The length is twice the width. What is the area of the rectangle?

   Perimeter: _____   Draw a picture:

   Length: _____

   Width: _____

   Area: _____

3. Melissa is 15 years older than Brent. Brent is 6 years older than Gael. If Brent is 9 years old, how old are Melissa and Gael?

   Write down what you know:

   Brent is _____ years old.   Brent is _____ years old, which is _____ years older than Gael.

   Melissa is _____ years older than Brent.

   To solve, _____ the numbers.   To solve, _____ the numbers.

   Melissa is _____ years old.   Gael is _____ years old.

Name _____

Reteaching 9
Lesson 95

- **Two-Step Problems about a Fraction of a Group**

  - Sometimes it takes two steps (operations) to find a fraction of a group when the total is known.
  - Divide by the **denominator** (bottom number) of the fraction to find the number in one part.
  - Multiply by the **numerator** (top number) of the fraction to find the number in more than one part.
  - Remember the rule we use to find a fraction of a group.

    Number in each group
    × Number of groups
    = Total

Example: At a concert, $\frac{7}{8}$ of the audience wore wristbands. If there were 5600 people in the audience, how many wore wristbands?

Solution: $8\overline{)5600} = 700$ divide the total by the denominator

$700 \times 7 = 4900$ multiply the result by the numerator

4900 people wore wristbands.

Practice:

1. $\frac{2}{3}$ of the 21 students wanted to play kickball. How many students wanted to play kickball?

   _____ students

2. Three-fifths of the 55 beads in Ms. Cavender's necklace were purple. How many beads were not purple?

   _____ beads were not purple.

104   © Harcourt Achieve Inc. and Stephen Hake. All rights reserved.   Saxon Math Intermediate 4

Name _____

Reteaching 96
Lesson 96

- **Average**

To find average:

1. Add the numbers.
2. Count how many numbers were added together.
3. Divide the sum by that number.

Example:

There are four buckets of water. The first bucket has 15 pints, the second has 8 pints, the third has 9 pints, and the fourth bucket has 20 pints. What was the average number of pints per bucket?

buckets: 1   2   3   4
          15 + 8 + 9 + 20 = 52

13 pints per bucket

$$4\overline{)52}^{\,13}$$

Practice:

1. Jerrel has a book to read for class. He read 35 pages the first day and 75 pages the second day. If he reads 45 pages on the third and fourth day, how many pages does he read on average per day?

   ```
     3 5
     7 5
     4 5   )
   + 4 5
   ```
   _____ pages per day

2. Anastacia plays on a basketball team. In seven games the points she scored were 23, 42, 19, 29, 35, 48, and 7. What is Anastacia's point average per game?

   ```
     2 3
     4 2
     1 9   )
     2 9
     3 5
     4 8
   +   7
   ```
   _____ points per game

Saxon Math *Intermediate 4*

Name _____

Reteaching
Lesson 97

• Mean, Median, Mode, and Range

- The **mean** is the average of a list of numbers.
- The **median** is the middle number when the numbers are arranged in order. If there is an even number of things in a list, the median is the average of the two middle numbers.
- The **mode** is the number that repeats most in the list.
- The **range** is the difference between the least and the greatest numbers.

Practice:

1. Find the mean, median, mode, and range of temperatures shown below.

    69°, 71°, 74°, 62°, 75°, 51°

    First arrange the temperatures in order:

    ____, ____, ____, ____, ____, ____

    Mean (average): _____   Median: _____

    Mode: _____   Range: _____

2. Find the mean, median, mode, and range of this set of data:

    29, 23, 30, 32, 25, 46, 18

    Arrange in order:

    ____, ____, ____, ____, ____, ____, ____

    Mean (average): _____   Median: _____

    Mode: _____   Range: _____

3. Find the median of this set of data.

    78, 81, 85, 77, 83, 90

    Median: _____

    Explain your answer.

    _____

106   © Harcourt Achieve Inc. and Stephen Hake. All rights reserved.   *Saxon Math* Intermediate 4

Name _____ **Reteaching 98**

Lesson 98

• Geometric Solids

- Geometric shapes that take up space are called **geometric solids**.
- This rectangular solid is made up of 2 layers of cubes with 6 cubes in each layer (2 × 6 = 12). The rectangular solid is made up of 12 small cubes.

- **Face** → Flat surface of a geometric solid
- **Edge** → Line segment where faces meet
- **Vertex** → Corner where edges meet

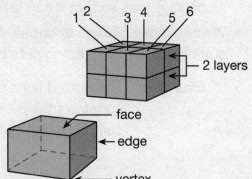

Practice:

For problems 1–4, name the shape of each object.

1. globe _____   2. tissue box _____

3. funnel _____   4. tuna can _____

5. What is the name of this solid? _____

6. How many edges does the figure above have? _____

7. Look at the figure of a rectangular solid.

   How many cubes in the top layer? _____

   How many layers? _____

   How many cubes altogether? _____

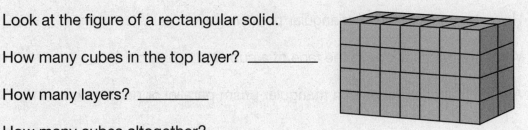

8. What type of solid has 8 vertices and 6 faces? _____
   Use the figures below to work out the answer.

Name _____

Reteaching

Lesson 99

- **Constructing Prisms**

  - A **net** is an arrangement of polygons (drawn on paper) that can be folded to become the faces of a geometric solid. Picture a cereal box that has been cut along its edges so that it can lay flat.
  - We can make models of cubes, rectangular prisms, and triangular prisms by cutting, folding and taping nets of these shapes.

  Example: The dotted lines on this net show where to fold it to construct the 3-dimensional model. What geometric solid does the net create?

  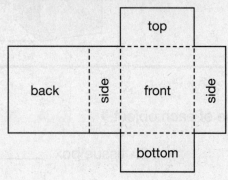

  If we fold the net on the dotted lines, we construct a rectangular prism.

  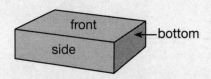

Practice:

1. How many faces does a triangular prism have? _____

2. What geometric shape is the face of a cube? _____

3. Are the triangular faces of a triangular prism parallel or perpendicular?

   _____

4. Look at the geometric solid below. Then sketch a net in the space at right.

   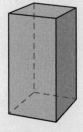

Name _____

Reteaching 100
Lesson 100

- ## Constructing Pyramids

  - Shapes that take up space are called **geometric solids**. Prisms like those we learned about in Lesson 98 are examples of geometric solids. **Pyramids** are another kind of geometric solid.

  - Geometric solids have three dimensions: **length**, **width**, and **height**.

Practice:

1. This pyramid has a square base. How many vertices does the pyramid have?

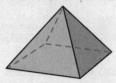

   _____ vertices

2. This pyramid has a triangular base. How many faces does the pyramid have?

   _____ faces

3. Which of these geometric solids is not a pyramid? _____

   A 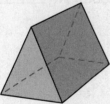   B   C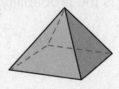

4. Look at the solid below. Then sketch a net of the solid.

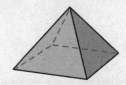

Saxon Math Intermediate 4   © Harcourt Achieve Inc. and Stephen Hake. All rights reserved.   109

Name _____

Reteaching
Investigation 10

- **Probability**

  - **Probability** measures how likely it is for an event to happen.
  - The **chance** of an event is often times expressed as a percent from 0% to 100%.
  - If an event is impossible, its probability is 0. Its chance is 0%.
  - All other events have probabilities between 0 and 1 or chances between 0% and 100%.

Practice:

1. What is the probability a tossed coin will land heads up? _____

2. What is the probability a number cube will stop on 3? _____

3. What is the probability a number cube will stop on a number less than 6?

   _____

4. What is the probability a number cube will stop on a number greater than 6?

   _____

5. If your team has a 30% chance of winning, is it more likely that you will win or not win? _____

6. If there is an 80% chance of rain, then what is the chance that it will not rain? Remember to write the percent symbol. _____

7. What is the probability that the arrow below will stop in sector A? _____

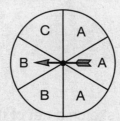

110 © Harcourt Achieve Inc. and Stephen Hake. All rights reserved. *Saxon Math Intermediate 4*

Name _____

Reteaching 101
Lesson 101

- **Tables and Schedules**

- **Tables** can be used to organize information.

Heights of Major Mountains

Mountain	Feet	Meters
Everest	29,035	8850
McKinley	20,320	6194
Kilimanjaro	19,340	5895
Matterhorn	14,691	4478
Pike's Peak	14,110	4301
Fuji	12,388	3776

Example: The Matterhorn is how many meters taller than Pike's Peak?

Solution: Use the numbers from the "Meters" column.

```
Matterhorn      4478 m
Pike's Peak   − 4301 m
                 177 m
```

- A **schedule** is a list of events organized by time.

School-Day Schedule

6:30 a.m.	Wake up, dress, eat breakfast	3:30 p.m.	Start homework
7:30 a.m.	Leave for school	5:00 p.m.	Play
8:00 a.m.	School starts	6:00 p.m.	Eat dinner
12:00 p.m.	Eat lunch	7:00 p.m.	Watch TV
2:45 p.m.	School ends, walk home	8:00 p.m.	Read
		8:30 p.m.	Shower
3:15 p.m.	Eat snack	9:00 p.m.	Go to bed

Practice:

Refer to the table and the schedule above. Remember to write the units.

1. The Matterhorn is how many feet higher than Pike's Peak? _____

2. Mt. McKinley is how many meters taller than Mt. Fuji? _____

3. At what time does the school day end? _____

4. At what time is dinner? _____

Saxon Math Intermediate 4 © Harcourt Achieve Inc. and Stephen Hake. All rights reserved. 111

Name _____

Reteaching 102
Lesson 102

- **Tenths and Hundredths on a Number Line**

  - On the number line below, the distance between whole numbers is divided into ten equal parts. The arrow is pointing to the number two and three tenths.

  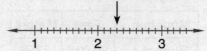

  - On the number line below, the distance between whole numbers is divided into ten equal parts. The arrow is pointing to the number 7 tenths. Notice that the mixed number is written in **decimal form** as 1.7.

  - The arrow below points to 6 plus a fraction. The fraction is 43 hundredths, which we can write as a decimal. The arrow points to 6.43.

  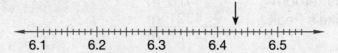

  - To round decimal numbers on a number line, find the decimal number, then find the nearest whole number. This can be a number to the right or left. If the decimal is exactly halfway between two whole numbers, use the number to the right (round up).

Practice:

Write the decimal number to which each arrow points:

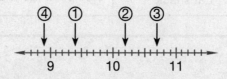

1. _____   2. _____   3. _____   4. _____

112

Saxon Math *Intermediate 4*

Name _____

Reteaching 103
Lesson 103

- **Fractions Equal to 1, and Fractions Equal to $\frac{1}{2}$**

  - Each of the circles below is divided into equal parts. The parts combine to equal one whole circle.
  - If the numerator (top number) and the denominator (bottom number) of a fraction are the same, the number equals 1.

$1 = \frac{2}{2}$ $1 = \frac{3}{3}$ $1 = \frac{4}{4}$ $1 = \frac{5}{5}$

  - If the numerator (top number) of a fraction is half the denominator (bottom number), then the fraction equals $\frac{1}{2}$.

$\frac{1}{2}$ $\frac{2}{4}$ $\frac{3}{6}$ $\frac{4}{8}$

Practice:

Use fraction manipulatives to help answer the questions.

1. Write the number 1 in fraction form with a denominator of 4. _____

2. Which of these fractions equals 1? _____

    A $\frac{5}{7}$ B $\frac{4}{7}$ C $\frac{8}{7}$ D $\frac{7}{7}$

3. Write a fraction equal to $\frac{1}{2}$ that has a denominator of 10. _____

4. Which of these fractions equals $\frac{1}{2}$? _____

    A $\frac{3}{4}$ B $\frac{1}{4}$ C $\frac{2}{4}$ D $\frac{4}{4}$

Saxon Math Intermediate 4

Name _____

Reteaching
Lesson 104

- **Changing Improper Fractions to Whole or Mixed Numbers**

  - To write an improper fraction as a mixed number, first divide the numerator by the denominator. The remainder can be written as the numerator of a fraction with the same denominator as the improper fraction.

    Example: Write $\frac{13}{3}$ as a mixed number.

    Solution: To find the number of wholes, divide.

    $3\overline{)13}$ = 4 R1 ← wholes / remainder

    The remainder becomes the numerator of the fractioned part of the mixed number.

    $\frac{13}{3} = 4\frac{1}{3}$

The picture below also shows that $\frac{13}{3}$ equals $4\frac{1}{3}$.

Practice:

Write each improper fraction as a mixed number. Then shade the circles to represent the mixed number.

1. $\frac{8}{3}$ = _____

2. $\frac{14}{4}$ = _____

3. $\frac{9}{2}$ = _____

4. $\frac{17}{5}$ = _____

5. $\frac{11}{6}$ = _____

Name _____

Reteaching 105
Lesson 105

- **Dividing by 10**

  - When dividing by 10, use **long division**.
  - There are four steps in long division: divide, multiply, subtract, and "bring down."
  - Remember:
     Use zero as a placeholder.
     Place a digit above each digit.

  Example: $10\overline{)537}$

  Solution: Ten will not divide into 5, but will divide into 53 five times. Be careful to write digits in the correct place above the dividend.
   1. Divide 53 by 10 and write "5."
   2. Multiply 5 by 10 and then write "50."
   3. Subtract 50 from 53 and write "3."
   4. Bring down the 7, making 37.

   ```
        5 3 R 7
   10)5 3 7
        5 0
           3 7
           3 0
              7
   ```

   Repeat steps:
   1. Divide 37 by 10 and write "3."
   2. Multiply 3 by 10 and then write "30."
   3. Subtract 30 from 37 and write "7."
   4. There is no number to bring down.

  - When dividing by 10, there will never be a remainder if the dividend ends in 0. If the dividend does not end in zero, the remainder will be the last digit of the dividend.

Practice:

1. $10\overline{)106}$ R _____

2. $10\overline{)329}$ R _____

3. $10\overline{)592}$ R _____

4. $10\overline{)1016}$ R _____

5. $10\overline{)780}$ _____

6. $10\overline{)860}$ _____

7. Which of these can be divided by 10 without a remainder? _____

   A 451   **B** 607   **C** 390   **D** 129

Name _____

Reteaching 10
Lesson 106

- **Evaluating Expressions**

  - **Evaluate** means find a value or answer a question.
  - These are expressions:

    $n + 8 \qquad t - 5 \qquad 4r$

  - The letter in an expression is called a **variable** because it can mean something different in each problem.
  - When you know what the letter represents (its value), you can solve the problem.

    Example: If n is 7, then what is the value of each of these expressions?

    $n + 5 \qquad n - 5 \qquad 2n$
    $7 + 5 = 12 \qquad 7 - 5 = 2 \qquad 2 \times 7 = 14$

Practice:

1. If q equals 14, then what is the value of $q - 9$? _____

2. Evaluate $s + t$ when $s = 9$ and $t = 3$. _____

3. What is the value of mn when m is 5 and n is 8? _____

4. What is the value of b^2 when b is 3? _____

5. If $a = lw$, then what is a when l is 10 and w is 6? _____

6. Evaluate $\frac{x}{y}$ using $x = 15$ and $y = 3$. _____

7. Find the value of $\sqrt{g}$ when g is 64. _____

8. What is the value of $x + 3x + 5x$ when x is 2? _____

Name _____

Reteaching 107

Lesson 107

- # Adding and Subtracting Fractions with Common Denominators

- When adding fractions with common denominators:

    Add only the numerators (top numbers).
    The denominator (bottom number) stays the same.

    $$\begin{array}{r} \frac{3}{7} \\ + \frac{3}{7} \\ \hline \frac{6}{7} \end{array}$$

- When subtracting fractions with common denominators:

    Subtract only the numerators (top numbers).
    The denominator (bottom number) stays the same.

    $$\begin{array}{r} \frac{3}{7} \\ - \frac{2}{7} \\ \hline \frac{1}{7} \end{array}$$

Practice:

1. $\frac{2}{3} + \frac{1}{3} = $ _____

2. $\frac{1}{4} + \frac{2}{4} = $ _____

3. $\frac{2}{5} + \frac{2}{5} = $ _____

4. $\frac{9}{11} - \frac{4}{11} = $ _____

5. $\frac{4}{5} - \frac{3}{5} = $ _____

6. $\frac{3}{4} - \frac{2}{4} = $ _____

7. $\frac{2}{6} + \frac{3}{6} = $ _____

8. $\frac{1}{4} + \frac{2}{4} = $ _____

9. $\frac{4}{5} - \frac{2}{5} = $ _____

Name _____

Reteaching 108

Lesson 108

- **Formulas**
- **Distributive Property**

- A **formula** is a rule for finding the answer to a problem.

    Area = length × width

- Formulas are usually written with letters for abbreviation.

    $A = lw$

- This chart shows common formulas.

    P represents the perimeter.
    s represents the length of a side of a square.
    b represents the base of a triangle or parallelogram.
    h represents the height of a triangle or parallelogram.

Common Formulas	
Area of a rectangle	$A = lw$
Area of a triangle	$A = \frac{bh}{2}$
Perimeter of a rectangle	$P = 2(l + w)$ $P = 2l + 2w$
Area of a square	$A = s^2$
Perimeter of a square	$P = 4s$
Area of a parallelogram	$A = bh$

- The formula for the perimeter of a rectangle demonstrates the **Distributive Property of Multiplication**.

    Perimeter of a rectangle $P = 2(l + w)$ and $P = 2l + 2w$

    $$2(l + w) = 2 \times l + 2 \times w$$

    Example: Use the Distributive Property to multiply:

    $$3(10 + 2)$$

    To use the Distributive Property, multiply first and then add the products.

    $$3(10 + 2) = (3 \times 10) + (3 \times 2)$$
    $$= 30 + 6$$
    $$= 36$$

Practice:

Use the Distributive Property to multiply:

1. $5(2 + 10) =$

    _____ × _____ + _____ × _____ = _____

2. $8(2 + 7) =$

    _____ × _____ + _____ × _____ = _____

3. $4(10 + 2) =$ _____

4. $3(20 + 3) =$ _____

Name _____

Reteaching 109
Lesson 109

- **Equivalent Fractions**

  - Equal portions of each circle below have been shaded. Notice that different fractions are used to name the shaded portions.

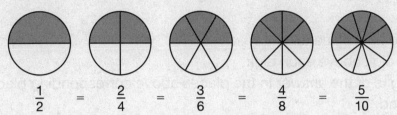

  - These fractions all name the same amount. Different fractions that name the same amount are called **equivalent fractions**.

Practice:

Name the equivalent fractions shown.

1.   _____ = _____

2. 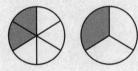   _____ = _____

Shade the rectangles to show that the following pairs of fractions are equivalent:

3.

4.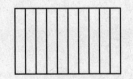

5.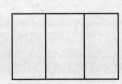

Name _____

Reteaching 110
Lesson 110

- **Dividing by Multiples of 10**

  - Use long division to divide by multiples of 10, just like dividing by 10.
    Divide, multiply, subtract, and "bring down."
  - Remember:
    Use zero as a placeholder.
    Place a digit above each digit.
  - Place the digits of the answer in the places above corresponding places in the dividend.

  Example:
  $$30 \overline{)88} \quad \begin{array}{r} 0\,2 \text{ R}28 \\ -60 \\ \hline 28 \end{array}$$

Practice:

Solve using long division.

1. $20\overline{)75}$ 0 R

2. $40\overline{)165}$ 0 0 R

3. $30\overline{)51}$ 0 R

4. $60\overline{)493}$ 0 0 R

5. $50\overline{)760}$ 0 R

6. $70\overline{)210}$ 0 0

Name _____

Reteaching Inv. 11
Investigation 11

- # Volume

  - The **volume** of a shape is the amount of space the shape occupies.
  - Volume is measured in **cubic units**.
  - $V = l \times w \times h$

  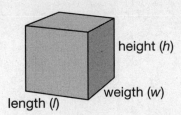

  Example: Find the volume of this rectangular solid.

  $V = l \times w \times h$

  $V = 3 \times 2 \times 2$

  $V = 12$ cubic centimeters

  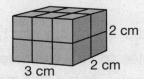

  - To estimate volume, round the measures before multiplying.

Practice:

Find the volume of each rectangular solid. Remember to write the units.

1.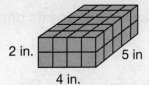

2.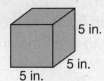

   _____   _____

3.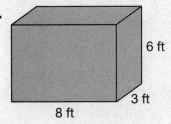

4.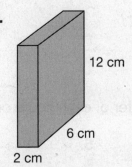

   _____   _____

Saxon Math *Intermediate 4*   © Harcourt Achieve Inc. and Stephen Hake. All rights reserved.   **121**

Name _____ **Reteaching 11**

Lesson 111

- **Estimating Perimeter, Area, and Volume**

  - We can use grids to estimate the areas and perimeters of shapes that are not regular polygons, or when we don't know the dimensions of the shape.

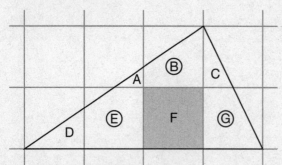

  - To estimate the area of the triangle, count the whole squares (F) and the mostly full squares (E, B, G).
  - Do not count squares that are mostly empty (D, A, C).
  - To estimate the perimeter, add the estimated lengths of the sides. The base is 4 units. We can see that one side is a little more than 3 units. The third side is a little more than 2 units. So, we estimate the perimeter to be between 9 units and 10 units.
  - One way to estimate the volume of a container is to fill the container with unit cubes and count the cubes.

Practice:

Estimate the area of each figure on these grids. Each small square represents one square inch. Remember to write the units.

1.    2. 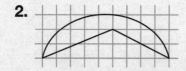   3.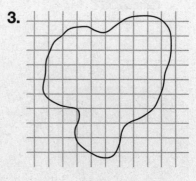

Estimate the perimeter of each figure on these grids.

4. 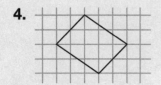   5.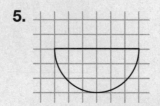

122   © Harcourt Achieve Inc. and Stephen Hake. All rights reserved.   *Saxon Math* Intermediate 4

Name _____

Reteaching 112
Lesson 112

- **Reducing Fractions**

- Use mental math to divide the numerator (top number) and the denominator (bottom number) by the same number.

   Example: Reduce $\frac{8}{12}$ (*Hint*: Divide 8 and 12 by 4)

   $$\frac{8}{12} = \frac{8 \div 4}{12 \div 4} = \frac{2}{3}$$

Practice:

Write the reduced form of each fraction. Some may already be reduced.

1. $\frac{2}{6} =$ _____
   Hint: ÷ 2

2. $\frac{3}{6} =$ _____
   Hint: ÷ 3

3. $\frac{4}{8} =$ _____
   Hint: ÷ 4

4. $\frac{6}{9} =$ _____

5. $\frac{2}{12} =$ _____

6. $\frac{4}{12} =$ _____

7. $\frac{4}{10} =$ _____

8. $\frac{10}{30} =$ _____

9. $\frac{7}{10} =$ _____

10. $\frac{6}{10} =$ _____

11. $\frac{6}{8} =$ _____

12. $\frac{6}{12} =$ _____

Reteaching 11

Lesson 113

- **Multiplying a Three-Digit Number by a Two-Digit Number**

- Use mental math to carry.
- Indent using 0 as a placeholder.

Practice:

Multiply. Remember to write the dollar sign and decimal point in money problems.

1.   256
  × 32
  _____
  +   0

2.   435
  × 18
  _____
  +   0

3. $1.67
  ×   23
  _____
  +   0

4.   511
  × 30
  _____
  +   0

5. $8.04
  ×   29
  _____
  +   0

6.   543
  × 21
  _____
  +   0

7. $4.92
  ×   36
  _____
  +   0

8.   714
  × 35
  _____
  +   0

Name _____

Reteaching 114

Lesson 114

- ## Simplifying Fraction Answers

- Change an improper fraction answer to a mixed number.
- Always **reduce** fraction answers if possible.
- Sometimes it helps to "break apart" a mixed number answer and change the improper fraction to another mixed number.

Example:

$$3\frac{4}{5}$$
$$+\ 4\frac{3}{5}$$
$$7\frac{7}{5} = 7 + 1\frac{2}{5} = 8\frac{2}{5}$$

Practice:

Simplify the answer to each sum or difference. (Change improper fractions to mixed numbers, and reduce if possible.)

1. $\frac{3}{5} + \frac{3}{5} =$ _____

2. $\frac{5}{8} - \frac{1}{8} =$ _____

3. $2\frac{3}{7}$
   $+\ 1\frac{6}{7}$

   $=$ _____

4. $6\frac{1}{4}$
   $+\ 3\frac{3}{4}$

   $=$ _____

5. $5\frac{5}{9}$
   $-\ 3\frac{2}{9}$

   $=$ _____

6. $4\frac{6}{10}$
   $+\ 7\frac{8}{10}$

   $=$ _____

Saxon Math Intermediate 4

Name _____

Reteaching 115
Lesson 115

- **Renaming Fractions**

  - Rename fractions using the loop method.
    Multiply the numbers in the loop and divide by the number outside the loop.

    Example:

    $\frac{3}{4} = \frac{?}{12}$ $\begin{array}{l} 12 \times 3 = 36 \\ 36 \div 4 = 9 \end{array}$ $\frac{3}{4} = \frac{9}{12}$

Practice:

Complete each equivalent fraction.

1. $\frac{1}{5} = \frac{}{15}$

2. $\frac{2}{3} = \frac{}{15}$

3. $\frac{4}{7} = \frac{}{14}$

4. $\frac{1}{2} = \frac{}{14}$

5. $\frac{5}{8} = \frac{}{24}$

6. $\frac{5}{12} = \frac{}{24}$

7. $\frac{4}{5} = \frac{}{30}$

8. $\frac{5}{6} = \frac{}{30}$

Name _____

Reteaching 116
Lesson 116

- **Common Denominators**

  - To rename fractions so that they have a **common denominator**:
    1. Look down the multiplication table column for each denominator.
    2. Find the smallest number both columns share.
    3. Rename. (Use the loop method.)

  Example: $\frac{3}{4}$ and $\frac{4}{5}$

  1. Go down the 4s column and the 5s column.
  2. Find that they both share a 20.
  3. Rename.

  $\frac{3}{4} = \frac{?}{20}$ $20 \times 3 = 60$ $\frac{15}{20}$
  $60 \div 4 = 15$

  $\frac{4}{5} = \frac{?}{20}$ $20 \times 4 = 80$ $\frac{16}{20}$
  $80 \div 5 = 16$

Practice:

1. Rename $\frac{1}{3}$ and $\frac{1}{7}$ so that they have a common denominator of 21.

   $\frac{1}{3} = \frac{\ }{21}$ $\frac{1}{7} = \frac{\ }{21}$

2. Rename $\frac{3}{5}$ and $\frac{5}{6}$ so that they have a common denominator of 30.

   $\frac{3}{5} = \frac{\ }{30}$ $\frac{5}{6} = \frac{\ }{30}$

Rename each pair of fractions using their least common denominator.

3. $\frac{2}{3} = \frac{\ }{\ }$ 4. $\frac{1}{4} = \frac{\ }{\ }$

   $\frac{3}{4} = \frac{\ }{\ }$ $\frac{3}{8} = \frac{\ }{\ }$

Saxon Math Intermediate 4   © Harcourt Achieve Inc. and Stephen Hake. All rights reserved.

Name _____

Reteaching 11
Lesson 117

• Rounding Whole Numbers Through Hundred Millions

• Review the Place Value chart.

Place Value

100,000,000	10,000,000	1,000,000	,	100,000	10,000	1000	,	100	10	1	.	$\frac{1}{10}$	$\frac{1}{100}$	$\frac{1}{1000}$
hundred millions	ten millions	millions		hundred thousands	ten thousands	thousands		hundreds	tens	ones	decimal point	tenths	hundredths	thousandths

• To round numbers through hundred millions:
  1. Underline the place value that you will be rounding to.
  2. Ask: Is the number to the right 5 or more? (5, 6, 7, 8, 9)
     Yes → Add 1 to the underlined number.
     No → The underlined number stays the same.
  4. Replace the numbers after the underlined number with zero.

  Examples:

  4<u>3</u>8,270 → 440,000   1<u>2</u>,876,250 → 13,000,000

  <u>4</u>27,063 → 400,000   <u>4</u>27,063,000 → 400,000,000

Practice:

Round each number to the nearest ten thousand.

1. 17,652 ___ ___ , ___ ___ ___   2. 31,984 ___ ___ , ___ ___ ___

Round each number to the nearest hundred thousand.

3. 747,985 ___ ___ ___ , ___ ___ ___   4. 658,103 ___ ___ ___ , ___ ___ ___

Round 4,586,712 to the nearest million.

5. ___ , ___ ___ ___ , ___ ___ ___

Round 46,029,984 to the nearest ten million.

6. ___ ___ , ___ ___ ___ , ___ ___ ___

Round 792,463,051 to the nearest hundred million.

7. ___ ___ ___ , ___ ___ ___ , ___ ___ ___

Name _____

Reteaching 118
Lesson 118

- **Dividing by Two-Digit Numbers**

  - Use zero as a placeholder.
  - When you multiply, if the answer is greater than the dividend, try a smaller number.
  - If the ones digit in the divisor is 4 or less, round it down before guessing.

    $$44\overline{)163}^{\,0\,0} \quad \text{(Think: } 40\overline{)160}^{\,0\,0}\text{)}$$

  - If the ones digit in the divisor is 5 or more, round it up before guessing.

    $$46\overline{)163} \quad \text{(Think: } 50\overline{)160}\text{)}$$

  - Then use the four division steps to solve (divide, multiply, subtract, bring down).

Practice:

Divide using long division.

1. $22\overline{)129}^{\,0\,0}$

2. $17\overline{)83}^{\,0}$

3. $57\overline{)243}^{\,0\,0}$

4. $72\overline{)284}^{\,0\,0}$

5. $51\overline{)299}^{\,0\,0}$

6. $14\overline{)63}^{\,0}$

7. $25\overline{)121}^{\,0\,0}$

8. $84\overline{)642}^{\,0\,0}$

Name _____

Reteaching 11
Lesson 119

- **Adding and Subtracting Fractions with Different Denominators**

- To add or subtract fractions that have different denominators, first rename the fractions so that they have common denominators.
    1. Find a common denominator.
    2. Rename. (Use the loop method.)
    3. Add or subtract the renamed fractions.

Example:

$$\begin{array}{r}\frac{2}{3} = \frac{}{9} \\ +\frac{4}{9} = \frac{}{9} \\ \hline \end{array} \qquad \begin{array}{r}\frac{2}{3} = \frac{6}{9} \\ +\frac{4}{9} = \frac{4}{9} \\ \hline \frac{10}{9} = 1\frac{1}{9}\end{array}$$

Practice:

Find each sum or difference.

1. $\begin{array}{r}\frac{1}{4} = \frac{}{8} \\ +\frac{3}{8} = \frac{}{8} \\ \hline \end{array}$

2. $\begin{array}{r}\frac{3}{4} = \frac{}{} \\ -\frac{1}{2} = \frac{}{} \\ \hline \end{array}$

3. $\begin{array}{r}\frac{2}{5} = \frac{}{} \\ +\frac{3}{10} = \frac{}{} \\ \hline \end{array}$

4. $\begin{array}{r}\frac{5}{8} = \frac{}{} \\ +\frac{1}{4} = \frac{}{} \\ \hline \end{array}$

5. $\begin{array}{r}\frac{4}{9} = \frac{}{} \\ +\frac{1}{3} = \frac{}{} \\ \hline \end{array}$

6. $\begin{array}{r}\frac{6}{9} = \frac{}{} \\ -\frac{2}{3} = \frac{}{} \\ \hline \end{array}$

7. $\begin{array}{r}\frac{9}{10} = \frac{}{} \\ -\frac{2}{5} = \frac{}{} \\ \hline \end{array}$

8. $\begin{array}{r}\frac{7}{8} = \frac{}{} \\ -\frac{1}{4} = \frac{}{} \\ \hline \end{array}$

Name _____

Reteaching 120
Lesson 120

- **Adding and Subtracting Mixed Numbers with Different Denominators**

- To add or subtract mixed numbers with different denominators:
    1. Copy the problem vertically.
    2. Rename the fractions so that they have common denominators.
    3. Add or subtract the fraction side.
    4. Add or subtract the whole numbers.
    5. Reduce the fraction side.

Example:
$$3\frac{1}{7} = 3\frac{2}{14}$$
$$+\ 2\frac{6}{14} = 2\frac{6}{14}$$
$$\overline{}$$
$$5\frac{8}{14} = 5\frac{4}{7}$$

Practice:

Find each sum or difference. Reduce when possible.

1. $2\frac{1}{3} + 1\frac{12}{15}$ $2\frac{1}{3} = 2\frac{}{}$

    $+\ 1\frac{12}{15} = 1\frac{12}{15}$

    _____

2. $5\frac{3}{7} + 2\frac{5}{14}$ $5\frac{3}{7} = \frac{}{}$

    $+\ \frac{}{} = \frac{}{}$

    _____

3. $4\frac{2}{3} + 3\frac{6}{9}$ $4\frac{2}{3} = \frac{}{}$

    $+\ \frac{}{} = \frac{}{}$

    _____

4. $6\frac{1}{8} + 3\frac{3}{4}$ $6\frac{1}{8} = \frac{}{}$

    $+\ \frac{}{} = \frac{}{}$

    _____

5. $3\frac{5}{8} - 2\frac{1}{4}$ $3\frac{5}{8} = 3\frac{5}{8}$

    $-\ 2\frac{1}{4} = 2\frac{}{}$

    _____

6. $2\frac{6}{10} - 2\frac{2}{5}$ $2\frac{6}{10} = \frac{}{}$

    $-\ \frac{}{} = \frac{}{}$

    _____

7. $9\frac{11}{12} - 7\frac{5}{6}$ $9\frac{11}{12} = \frac{}{}$

    $-\ \frac{}{} = \frac{}{}$

    _____

8. $5\frac{9}{14} - 4\frac{3}{7}$ $5\frac{9}{14} = \frac{}{}$

    $-\ \frac{}{} = \frac{}{}$

    _____

Saxon Math Intermediate 4

Name _____

Reteaching Inv.
Investigation 12

- **Solving Balanced Equations**

  - An **equation** states two quantities are equal.
  - A model for an equation is a balance scale.
  - All equations have an equal sign.
  - To balance and solve the equation we must follow the steps below:
    – Read the problem.
    – Identify type of problem (combining, separating).
    – Get the block with the letter by itself on one side. Remember what you do to one side, you have to do the other.
    – Find the number that solves the equation.
    – Check your answer.

Practice:

1.

   Equation: _____

   Solution: N = _____

2.

   Equation: _____

   Solution: N = _____

3.

   Equation: _____

   Solution: N = _____

4.

   Equation: _____

   Solution: N = _____

5.

   Equation: _____

   Solution: N = _____

6.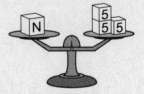

   Equation: _____

   Solution: N = _____

Answer Key

Answers

Reteaching 1

1. 9
2. 3
3. 11
4. 3 + 8 = 11; 8 + 3 = 11
5. $N = 5$
6. $N = 8$

Reteaching 2

1. 4; $n = 4$
2. 7; $n = 7$
3. 6; $y = 6$
4. 29; 5; $n = 5$
5. 35
6. 32
7. 45

Reteaching 3

1. 6, 5; Count down by 1
2. 14, 17; Count up by 3
3. 20
4. 7
5. Three
6. Four
7. 4
8. 1

Reteaching 4

1. 1 hundreds, 3 tens, 2 ones
2. 3 hundreds, 2 tens, 4 ones; $132
3. Tens; Ones; Hundreds
4. 683

Reteaching 5

1. Third; Ninth
2. See student work.
3. 1/16/any year
4. 12
5. May
6. August
7. February
8. 9/21/07
9. See student work.

Reteaching 6

1. 9; 8 + 9 = 17
2. 5; 7 + 5 = 12
3. 8; 6 + 8 = 14
4. 9; 9 + 9 = 18
5. 7; 4 + 7 = 11
6. 7; 8 + 7 = 15
7. Adding; see student work.

Reteaching 7

1. Three
2. Seventy-five
3. Eighty-eight
4. Three hundred sixty-seven
5. Six hundred tweny-nine
6. 15
7. 37
8. 107
9. 362

Answers

Reteaching 8

1. $57
2. $43
3. $96
4. $89
5. $94
6. $85
7. $94
8. $97

Reteaching 9

1. $91
2. $62
3. $102
4. $81
5. $90
6. $100
7. $88
8. $141
9. $81

Reteaching 10

1. Even
2. Odd
3. Even
4. Odd
5. Odd
6. Even
7. 850, 852, 854, 856, 858
8. 491, 493, 495, 497, 499
9. See student work.

10. See student work.

Investigation 1

1. 35
2. −5
3. >
4. <
5. >
6. See student work.

Reteaching 11

1. 7
2. 9
3. 13 − 5
4. 12 − 8
5. 8
6. 6

Reteaching 12

1. 6
2. 11
3. 9
4. 15
5. See student work.
6. 8
7. 4
8. 15
9. 3

Reteaching 13

1. $514
2. $927
3. 1059

Saxon Math Intermediate 4 © Harcourt Achieve Inc. and Stephen Hake. All rights reserved. 135

Answers

4. 660
5. $935
6. 1009
7. $1087
8. $1000
9. 300

Reteaching 14

1. $114
2. $464
3. 250
4. 265
5. $111
6. 222
7. $M = 17$
8. $X = 33$

Reteaching 15

1. $126
2. $417
3. 207
4. 157
5. $109
6. 219
7. 208
8. 439
9. 27

Reteaching 16

1. 700 + 60 + 4
2. 500 + 10 + 9
3. 400 + 6

4. 600 + 10
5. $W = 11$
6. $P = 33$
7. $N = 89$
8. $X = 17$

Reteaching 17

1. 159
2. 159
3. 277
4. 223
5. 252
6. 229

Reteaching 18

1. 98°F
2. 16°C
3. 82°F; see student work.

Reteaching 19

1. 4:30 p.m.
2. 11:13 a.m.
3. 6:41 a.m.
4. 6:50
5. 3:25

Reteaching 20

1. 80
2. 30
3. 60
4. $9.00
5. $6.00

Answers

6. $4.00

7. $11.00

8. $18.00

9. $30.00

Investigation 2

1. Length = 3 inches; width = 2 inches
2. 10 inches
3. 8 centimeters
4. 20 inches

Reteaching 21

1. See student work.
2. See student work.
3. See student work.
4. See student work.

Reteaching 22

1. $\frac{3}{4}$
2. $\frac{3}{5}$
3. $\frac{1}{6}$
4. $\frac{5}{8}$
5. $\frac{4}{9}$
6. $\frac{1}{4}$
7. $3.91
8. $8.70

Reteaching 23

1. Sample:
2. Sample:
3. See student work.

Reteaching 24

1. 17
2. 14
3. 24
4. 13
5. 37
6. 19
7. 38
8. 14
9. 23
10. 65
11. 90

Reteaching 25

1. 25
2. 83
3. $16

Reteaching 26

1. Sample:
2. Sample:
3. Sample:
4. Sample:
5. No; Sample: Fifths are equal parts; this is not an equal part.

Answers

Reteaching 27

1. 2 × 7
2. 8 × 4
3. 6 + 6 + 6
4. 4 + 4 + 4 + 4 + 4
5. 3 + 3 + 3 + 3 + 3 + 3 + 3 + 3
6. 2:30 p.m.
7. 1:00 p.m.

Reteaching 28

1. 54
2. 54
3. 32
4. 32
5. 56
6. 56
7. 45
8. 45

Reteaching 29

1. 15
2. 16
3. 0
4. 7
5. 35
6. 6
7. 5
8. 4
9. 12
10. 20
11. 45
12. 4
13. 6
14. 12
15. 0
16. 0
17. 30
18. 14
19. 25
20. 2
21. 10

Reteaching 30

1. $174
2. $1.99
3. 187
4. 183
5. 150
6. 287

Investigation 3

1. See student work.
2. See student work.
3. See student work.
4. 8 × 3 = 24
5. 6
6. 9
7. 10

Reteaching 31

1. 6
2. 15
3. 14

Answers

Reteaching 32

1. 63
2. 90
3. 18
4. 60
5. 44
6. 54
7. 36
8. 99

Reteaching 33

1. Two hundred eighteen thousand
2. Five hundred sixteen thousand
3. Sixteen thousand, three hundred
4. Six hundred, seventy-five thousand
5. Six thousand, three hundred eighty-five
6. Four hundred fifty thousand, two hundred ninety-five
7. Three thousand, one hundred twelve
8. Twenty-one thousand, two hundred eighty three
9. Six hundred eighteen thousand, four hundred ninety-three

Reteaching 34

1. 2,215,685
2. 8,045,815
3. 401,976,009
4. 8,711,256
5. 563,000,008

Reteaching 35

1. $2\frac{1}{3}$
2. $1\frac{6}{8}$ or $1\frac{3}{4}$
3. Ten and three fourths
4. Three and seventy-five hundredths
5. 34¢
6. 71¢
7. $0.63
8. $0.05

Reteaching 36

1. $0.60
2. $\frac{6}{10}$
3. $\frac{3}{10}$
4. $0.15
5. $\frac{45}{100}$
6. $0.45
7. <

Reteaching 37

1. $2\frac{1}{3}$
2. $\frac{3}{4}$
3. $\frac{2}{4}$ or $\frac{1}{2}$
4. $22\frac{1}{6}$
5. $10\frac{3}{8}$
6. $1\frac{5}{6}$

Reteaching 38

1. $3 \times 6 = 18$; $6 \times 3 = 18$; $\frac{18}{3} = 6$; $\frac{18}{6} = 3$
2. $8 \times 4 = 32$; $4 \times 8 = 32$; $\frac{32}{4} = 8$, $\frac{32}{8} = 4$
3. $8 \times 7 = 56$; $7 \times 8 = 56$; $\frac{56}{8} = 7$; $\frac{56}{7} = 8$
4. $8 \times 6 = 48$; $6 \times 8 = 48$; $\frac{48}{8} = 6$; $\frac{48}{6} = 8$
5. 12
6. 24

Answers

7. 28
8. 18
9. 21
10. 24
11. 24
12. 30
13. 42
14. 56
15. 56
16. 32

Reteaching 39

1. $\frac{1}{2}$
2. $1\frac{3}{4}$
3. $2\frac{1}{4}$
4. $3\frac{7}{8}$
5. $\frac{3}{4}$
6. $1\frac{1}{4}$
7. $2\frac{2}{4}$
8. 4

Reteaching 40

1. 4
2. 8
3. 16
4. 1 liter
5. 1000
6. 5000
7. 2 c; 2 p; 4 qt

Investigation 4

1. See student work.

2. See student work.
3. Sixty-three hundredths
4. Sixty-three hundredths
5. Four and eight tenths
6. Fifteen and ninety-three hundredths
7. 6.7
8. 8.29

Reteaching 41

1. $175
2. $0.39
3. 179
4. 9
5. 3
6. 7
7. 12

Reteaching 42

1. 600
2. 600
3. 100
4. 1700
5. 4100
6. 100
7. 900
8. 300
9. 900

Reteaching 43

1. $8.12
2. $4.17
3. $1.11

Answers

4. 6.81
5. 1.59
6. 1.87
7. 3.04
8. 0.19
9. 7.6

Reteaching 44

1. 96
2. 136
3. 129
4. 220
5. 255
6. 108
7. 248
8. 128

Reteaching 45

1. 0
2. 10
3. 15
4. 21
5. 20
6. 32

Reteaching 46

1. 5
2. 4
3. 6
4. 6
5. 2
6. 4

7. 6
8. 7
9. $12 \div 3 = 4; 3 \times 4 = 12; 4 \times 3 = 12$

Reteaching 47

1. 8
2. 5
3. 4
4. 5
5. 3
6. 6
7. See student work.
8. See student work.
9. $7 \times 6 = 42; 6 \times 7 = 42; \frac{42}{7} = 6; \frac{42}{6} = 7$

Reteaching 48

1. 68
2. 208
3. $175
4. 153
5. $576
6. 148
7. 520
8. $441
9. 476

Reteaching 49

1. 42
2. 36
3. 20

Saxon Math Intermediate 4 © Harcourt Achieve Inc. and Stephen Hake. All rights reserved. 141

Answers

Reteaching 50

1. 6
2. 3
3. 2
4. 6.96
5. 1.36
6. 18.57

Investigation 5

1. 25%
2. >
3. <
4. D

Reteaching 51

1. 1841
2. 68, 617
3. 725, 156
4. 8
5. 7
6. 9
7. 3

Reteaching 52

1. 627
2. 25,421
3. 86,202
4. 8
5. 9

Reteaching 53

1. See student work; 4 R 2
2. 7 R 1
3. 2 R 3
4. 5 R 1
5. 8 R 1
6. 4 R 3
7. 6 R 2

Reteaching 54

1. Every 4 years
2. 05, 21, 2014
3. 45
4. 8000
5. 3000

Reteaching 55

1. 1, 2, 3, 5, 7, 11
2. 24
3. even
4. 1, 2, 3, 4, 6, 12
5. 1, 2, 4, 8, 16
6. 2, 4, 5, 10
7. 1, 11
8. 1, 2, 3, 4, 6, 8, 12, 24

Reteaching 56

1. <
2. <
3. >
4. <
5. >
6. <

Answers

Reteaching 57

1. 455
2. 420
3. 45

Reteaching 58

1. 992
2. $3090
3. $27.09
4. 3006
5. $1053
6. $25.02
7. $30.73
8. 5784
9. 3664

Reteaching 59

1. 180; 176
2. 900; 873
3. 200; 223
4. 0; 5
5. 280; 273
6. 4000; 3720
7. 15; 15 R 3
8. 10; 9 R 1
9. Less than; 6000

Reteaching 60

1. 3 minutes
2. 8 seconds
3. 6 hours

Investigation 6

1. 3
2. 12 dogs
3. 6 Terriers
4. See student work.
5. See student work.

Reteaching 61

1. $\frac{5}{12}$
2. $\frac{3}{5}$
3. 3
4. 6
5. 2
6. 8

Reteaching 62

1. 48
2. 160
3. 36
4. 125
5. 28
6. 35
7. 3^5
8. 5^7

Reteaching 63

1. See student work.
2. See student work.
3. See student work.

Reteaching 64

1. 27
2. 13

Saxon Math Intermediate 4 © Harcourt Achieve Inc. and Stephen Hake. All rights reserved.

143

Answers

3. 25
4. 13
5. 14
6. 12
7. 27
8. 19
9. B; 7 and 2

Reteaching 65

1. 7
2. 4
3. 28
4. 39
5. 26
6. 51
7. 68

Reteaching 66

1. All three triangles are similar.
2. Triangles *XYZ* and *UVW*
3. Triangles *A* and *C*, and triangles *B* and *D*
4. Triangles *B* and *D*

Reteaching 67

1. 650
2. 410
3. $7.80
4. 1140
5. $87.90
6. 2960
7. $16 \times 4 \times 10 = 640$

Reteaching 68

1. 34 R 2
2. 33 R 6
3. 34 R 2
4. 43 R 3
5. 39 R 2
6. 49 R 4
7. 71 R 1
8. 76
9. 33 R 1
10. Mulitply; 4; add; 138

Reteaching 69

1. 300 cm
2. 50 dimes
3. 12 cm
4. 2.3 cm
5. 180 mm

Reteaching 70

1. 9
2. 7
3. 12
4. 14

Investigation 7

1. 10 students
2. Ten more black shoes
3. Black shoes
4. White shoes and blue shoes
5. Shoe colors other than white, blue, and black
6. See student work.

Answers

Reteaching 71
1. 40 R 2
2. 40 R 4
3. 30 R 5
4. 80 R 2
5. 90
6. 20 R 3
7. 30 R 2
8. 50 R 4

Reteaching 72
1. 9 hours
2. about $\frac{1}{2}$ an hour
3. 20 miles
4. $57
5. $25 more
6. $133
7. 16 hours
8. 5 hours
9. 21 hours

Reteaching 73
1. Rotation
2. Translation
3. Reflection

Reteaching 74
1. $\frac{5}{9}$
2. $\frac{8}{15}$
3. $\frac{3}{12}$ or $\frac{1}{4}$
4. $\frac{8}{14}$ or $\frac{4}{7}$
5. $\frac{13}{29}$

6. $\frac{5}{12}$
7. $\frac{29}{61}$
8. $\frac{9}{17}$

Reteaching 75
1. 180°
2. Clockwise; about 90°

Reteaching 76
1. 241 R 3
2. $1.73
3. 302 R 2
4. $8.75
5. 850 R 2
6. 66 R 2
7. 555 R 1
8. 1169 R 3

Reteaching 77
1. 6
2. 112
3. B
4. 100 grams

Reteaching 78
1. No. A triangle must have 3 angles and 3 sides.
2. Scalene
3. Equilateral
4. Obtuse
5. Right
6. 24 inches
7. No because all three sides have different lengths, and we only know the length of one side.

Saxon Math Intermediate 4

Answers

8. The sides have the same length.

Reteaching 79

1.

2.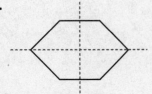

3.

4.

5. None

6.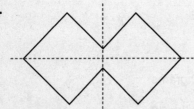

7. None

8. None

9.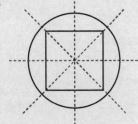

Reteaching 80

1. 180 R 2
2. 2006 R 1
3. 667 R 3
4. 75 R 3
5. 824 R 2
6. 331 R 5
7. 600
8. 900
9. 15,000

Investigation 8

1. See student work.
2. Saturday
3. Sunday
4. Highest temperature
5. Lowest temperature
6. 3°F
7. 91°F

Reteaching 81

1. 45°
2. 60°
3. 120°
4. 90°
5. 20°
6. 135°

Reteaching 82

1. See student work.

Answers

Reteaching 83

1. $7.06
2. $63.81
3. $16.19

Reteaching 84

1. 0.617
2. 5.346
3. 0.72
4. 2.084
5. $\frac{349}{1000}$; Three hundred forty-nine thousandths
6. $5\frac{128}{1000}$; Five and one hundred twenty-eight thousandths
7. $\frac{4}{1000}$; Four thousandths
8. $4\frac{405}{1000}$; Four and four hundred five thousandths

Reteaching 85

1. 4250
2. $53.00
3. $67.00
4. $375.00
5. 6000
6. $3220.00
7. 540
8. 41,800

Reteaching 86

1. 5600
2. $20.00
3. 12,000
4. $2700.00
5. 30,000

6. $12,000
7. 1000
8. 200,000

Reteaching 87

1. 864
2. 864
3. 1105
4. 1148

Reteaching 88

1. 12
2. 1
3. 1
4. 13
5. 26
6. 2
7. No

Reteaching 89

1.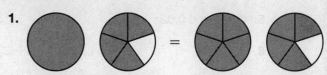
2.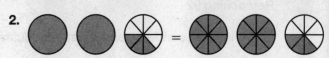
3.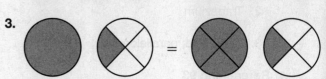
4.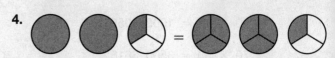

Reteaching 90

1. 1083
2. 2173

Answers

3. 5695
4. 2924
5. 1980
6. 432

Investigation 9

1. One-fifth
2. 2; see student work.
3. 100%, 50%
4. $\frac{1}{8}, \frac{1}{4}, \frac{1}{2}$
5. 25%
6. 0.25
7. $\frac{2}{4} = \frac{1}{2}$; see student work.

Reteaching 91

1. 8
2. 1
3. Hundredths
4. 78.95 and 78.950
5. 0.020 and 0.02
6. 5

Reteaching 92

1. Rhombus; parallelogram; rectangle; square
2. Trapezium
3. Rectangle; parallelogram

Reteaching 93

1. 50 × 20 = 1000; 1056
2. 70 × 50 = 3500; 3618
3. 30 × 20 = 600; 464
4. 60 × 30 = 1800; 1947

Reteaching 94

1. 50¢
2. 12 in.; 4 in.; 2 in.; 8 sq. in.; see student drawings.
3. 24; 3

Reteaching 95

1. 14
2. 22

Reteaching 96

1. 50
2. 29

Reteaching 97

1. 51, 62, 69, 71, 74, 75; 67; 70; none; 24
2. 18, 23, 25, 29, 30, 32, 46; 29; 29; none; 28
3. 82; I found the average of the two middle numbers.

Reteaching 98

1. Sphere
2. Rectangular prism
3. Cone
4. Cylinder
5. Pyramid
6. 8
7. 14, 4, 56
8. Rectangular prisms and cubes

Reteaching 99

1. 5
2. square
3. parallel
4. See student work.

Answers

Reteaching 100

1. 5
2. 4
3. A
4. See student work.

Investigation 10

1. $\frac{1}{2}$; 50%
2. $\frac{1}{6}$
3. $\frac{5}{6}$
4. 0
5. Not win
6. 20%
7. $\frac{3}{6} = \frac{1}{2}$

Reteaching 101

1. 581 feet
2. 2418 meters
3. 2:45 p.m.
4. 6:00 p.m.

Reteaching 102

1. 9.4
2. 10.2
3. 10.7
4. 8.9

Reteaching 103

1. $\frac{4}{4}$
2. D
3. $\frac{5}{10}$
4. C

Reteaching 104

1.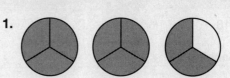
2.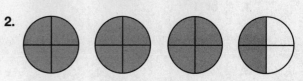
3.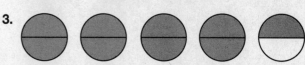
4.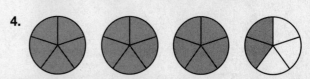
5.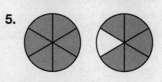

Reteaching 105

1. 10 R 6
2. 32 R 9
3. 59 R 2
4. 101 R 6
5. 78
6. 86
7. C

Reteaching 106

1. 5
2. 12
3. 40
4. 9
5. 60
6. 5

Answers

7. 8

8. 18

Reteaching 107

1. $\frac{3}{3}$ or 1
2. $\frac{3}{4}$
3. $\frac{4}{5}$
4. $\frac{5}{11}$
5. $\frac{1}{5}$
6. $\frac{1}{4}$
7. $\frac{5}{6}$
8. $\frac{3}{4}$
9. $\frac{2}{5}$

Reteaching 108

1. $5 \times 2 + 5 \times 10 = 60$
2. $8 \times 2 + 8 \times 7 = 72$
3. 48
4. 69

Reteaching 109

1. $\frac{6}{8} = \frac{3}{4}$
2. $\frac{2}{6} = \frac{1}{3}$
3.
4.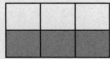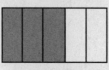
5.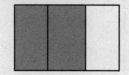

(Note: item 3 uses , with an adjacent figure; item 4 uses .)

Reteaching 110

1. 3 R 15
2. 4 R 5
3. 1 R 21
4. 8 R 13
5. 15 R 10
6. 3

Investigation 11

1. 40 cu. in.
2. 125 cu. in.
3. 144 cu. ft
4. 144 cu. ft
5. 24 boxes

Reteaching 111

1. About 7 sq. in.
2. About 12 sq. in.
3. About 10 sq. in.
4. About 14 sq. in.
5. About 91 sq. in.

Reteaching 112

1. $\frac{1}{3}$
2. $\frac{1}{2}$
3. $\frac{1}{2}$
4. $\frac{2}{3}$
5. $\frac{1}{6}$
6. $\frac{1}{3}$
7. $\frac{2}{5}$
8. $\frac{1}{3}$
9. $\frac{7}{10}$

10. $\frac{3}{5}$

11. $\frac{3}{4}$

12. $\frac{1}{2}$

Reteaching 113

1. 8192
2. 7830
3. $38.41
4. 15,330
5. $233.16
6. 11,403
7. $177.12
8. 24,990

Reteaching 114

1. $\frac{6}{5} = 1\frac{1}{5}$
2. $\frac{4}{8} = \frac{1}{2}$
3. $3\frac{9}{7} = 4\frac{2}{7}$
4. $9\frac{4}{4} = 10$
5. $2\frac{3}{9} = 2\frac{1}{3}$
6. $11\frac{14}{10} = 12\frac{4}{10} = 12\frac{2}{5}$

Reteaching 115

1. $\frac{3}{15}$
2. $\frac{10}{15}$
3. $\frac{8}{14}$
4. $\frac{7}{14}$
5. $\frac{15}{24}$
6. $\frac{10}{24}$
7. $\frac{24}{30}$
8. $\frac{25}{30}$

Reteaching 116

1. $\frac{7}{21}, \frac{3}{21}$
2. $\frac{18}{30}, \frac{25}{30}$
3. $\frac{8}{12}, \frac{9}{12}$
4. $\frac{2}{8}, \frac{3}{8}$

Reteaching 117

1. 20,000
2. 30,000
3. 700,000
4. 700,000
5. 5,000,000
6. 50,000,000
7. 800,000,000

Reteaching 118

1. 5 R 19
2. 4 R 15
3. 4 R 15
4. 3 R 68
5. 5 R 44
6. 4 R 7
7. 4 R 21
8. 7 R 54

Reteaching 119

1. $\frac{2}{8} + \frac{3}{8} = \frac{5}{8}$
2. $\frac{3}{4} - \frac{2}{4} = \frac{1}{4}$
3. $\frac{4}{10} + \frac{3}{10} = \frac{7}{10}$
4. $\frac{5}{8} + \frac{2}{8} = \frac{7}{8}$
5. $\frac{4}{9} + \frac{3}{9} = \frac{7}{9}$
6. $\frac{6}{9} - \frac{6}{9} = 0$

Answers

7. $\frac{9}{10} - \frac{4}{10} = \frac{5}{10}$

8. $\frac{7}{8} - \frac{2}{8} = \frac{5}{8}$

Reteaching 120

1. $4\frac{2}{15}$
2. $7\frac{11}{14}$
3. $8\frac{3}{9} = 8\frac{1}{3}$
4. $9\frac{7}{8}$
5. $1\frac{3}{8}$
6. $\frac{2}{10} = \frac{1}{5}$
7. $2\frac{1}{12}$
8. $1\frac{3}{14}$

Investigation 12

1. $n + 5 = 10; n = 5$
2. $6 = n + 6; n = 0$
3. $n + 5 = 9; n = 4$
4. $12 = n + 8; n = 4$
5. $n = 16$
6. $n = 15$